Stan Dolan

Discrete Mathematics For AQA

2

PUBLISHED BY THE PRESS SYNDICATE OF THE UNIVERSITY OF CAMBRIDGE
The Pitt Building, Trumpington Street, Cambridge, United Kingdom

CAMBRIDGE UNIVERSITY PRESS
The Edinburgh Building, Cambridge CB2 2RU, UK
40 West 20th Street, New York, NY 10011-4211, USA
10 Stamford Road, Oakleigh, VIC 3166, Australia
Ruiz de Alarcón 13, 28014 Madrid, Spain
Dock House, The Waterfront, Cape Town 8001, South Africa

http://www.cambridge.org

© Cambridge University Press 2001

This book is in copyright. Subject to statutory exception and to the provisions of relevant collective licensing agreements, no reproduction of any part may take place without the written permission of Cambridge University Press.

First published 2001

Printed in the United Kingdom at the University Press, Cambridge

Typefaces Times, Helvetica   *Systems* Microsoft® Word, MathType™

A catalogue record for this book is available from the British Library

ISBN 0 521 00655 4 paperback

The author and publisher would like to thank J. Sainsbury plc for permission to reproduce the diagram which appears on page 30.

Contents

	Introduction	iv
1	Allocation	1
2	Network flows	15
3	Critical path analysis	30
4	Dynamic programming	53
5	The Simplex algorithm	64
6	Game theory	76
	Revision exercise	95
	Mock examinations	100
	Answers	106
	Glossary	121
	Summary of algorithms	122
	Index	123

Introduction

This book has been written for the Discrete Mathematics module, D2, of AQA Specification A.

The book is divided into chapters roughly corresponding to specification headings. Occasionally a section includes an important result that is difficult to prove or outside the specification. These sections are marked with an asterisk (*) in the section heading, and there is usually a sentence early on explaining precisely what it is that the student needs to know.

It is important to recognise that, while every effort has been made by the author and by AQA to make the books match the specification, the books do not and must not define the examination. It is conceivable that questions might be asked in the examination, examples of which do not appear specifically in the book.

Occasionally within the text paragraphs appear in *this type style*. These paragraphs are usually outside the main stream of the mathematical argument, but may help to give insight, or suggest extra work or different approaches.

Numerical work is presented in a form intended to discourage premature approximation. In ongoing calculations inexact numbers appear in decimal form like 3.456..., signifying that the number is held in a calculator to more places than are given. Numbers are not rounded at this stage; the full display could be either 3.456 123 or 3.456 789. Final answers are then stated with some indication that they are approximate, for example '1.23 correct to 3 significant figures'.

There are plenty of exercises, and each chapter contains a Miscellaneous exercise which includes examination questions. Most of these questions were set in OCR examinations, and some were set in AQA examinations. Questions which go beyond examination requirements are marked by an asterisk. At the end of the book there is a set of Revision exercises and two practice examination papers. The author thanks Jan Dangerfield, who contributed to the exercises, and Hugh Neill, both of whom read the book very carefully and made many extremely useful and constructive comments.

The author thanks AQA and Cambridge University Press for their help in producing this book.

1 Allocation

This chapter is about matching the elements of one set with elements of another. When you have completed it you should

- be able to interpret allocation problems as matching problems where cost must be minimised
- know how to use the Hungarian algorithm to solve allocation problems.

1.1 Introduction

Matching the elements of two different sets is a common task which you often perform without being aware of it. When you distribute copies of a newsletter to your friends you are matching the set of newsletters with the set of friends. In this case the task is easy because any copy can be assigned to any friend.

Matching becomes more difficult when the two sets to be matched have characteristics that make certain assignments undesirable or impossible. For example, in matching people with jobs, the skills of the people and the requirements of the jobs mean that some assignments should not be made.

You have already met this kind of matching problem in D1 Chapter 5. In this chapter you will study another type of matching problem in which each assignment has a cost which you must try to minimise.

Suppose that a building company has four contracts which must be completed at the same time. The work is to be done by subcontractors, each of whom can carry out only one contract. The subcontractors' quotes for each of the four jobs are shown in Table 1.1.

		\multicolumn{4}{c}{Contract}			
		A	B	C	D
Subcontractor	1	10	5	9	–
	2	9	6	9	6
	3	10	–	10	7
	4	9	5	9	8

Quotes are in £1000s.
'–' indicates no quote.

Table 1.1

Here there is no difficulty in finding a matching. An example is 1-A, 2-B, 3-C, 4-D. The problem is to find a matching which minimises the total cost; such a problem is called an **allocation problem**.

A good method of tackling allocation problems is to reduce the array of costs by subtracting from each element of a row (or column) the least element in that row (or column). For example, Table 1.3 is obtained from Table 1.2 by subtracting 5, 6, 7 and 5 from the first, second, third and fourth rows respectively. The smallest number in each new row is now equal to zero.

10	5	9	–
9	6	9	6
10	–	10	7
9	5	9	8

$-5 \to$
$-6 \to$
$-7 \to$
$-5 \to$

5	0	4	–
3	0	3	0
3	–	3	0
4	0	4	3

Table 1.2

Table 1.3

The solution to the original problem will then simply be the solution to the new problem with an extra cost of £23,000 because $5 + 6 + 7 + 5 = 23$. Reducing the columns of Table 1.3 in a similar way, in this case by subtracting 3 from the first and third columns, leads to the array in Table 1.4. (The extra cost to be added to a solution is now £29,000.)

2	0	1	–
0	0	0	0
0	–	0	0
1	0	1	3

Table 1.4

For this matrix, it is easy to see that there are no matchings of zero cost but there are several of cost only 1. For example, Table 1.5 shows a matching of cost 1 in bold type. So the original problem has an optimal allocation, of cost £30,000, shown in Table 1.6.

2	**0**	1	–
0	0	**0**	0
0	–	0	**0**
1	0	1	3

Table 1.5

10	**5**	9	–
9	6	**9**	6
10	–	10	**7**
9	5	9	8

Table 1.6

In general, if you obtain one array from another by adding or subtracting *any* fixed number from each element of a row (or column), a solution to the new allocation problem gives a solution to the original problem, and vice versa. In Table 1.2, for example, adding 10 to each element of the first column corresponds to all the contractors raising their quotes on contract A by £10,000. The costs of all possible matchings go up by £10,000, so your best choice of matching remains the same.

> To solve an allocation problem, first reduce the array of costs by subtracting the least number in a row (or column) from each element of that row (or column).

Example 1.1.1
Choose five numbers from the array given below so that the sum of the five numbers is the least possible. No two numbers can be in the same row or column.

10	10	9	8	10
10	12	12	9	13
16	16	14	12	15
14	15	12	12	16
15	16	14	13	14

Reducing the array by rows leads to the array on the left; then reducing by columns leads to the array on the right, in which the minimum allocation is shown in bold type.

2	2	1	0	2
1	3	3	0	4
4	4	2	0	3
2	3	0	0	4
2	3	1	0	1

1	**0**	1	0	1
0	1	3	0	3
3	2	2	**0**	2
1	1	**0**	0	3
1	1	1	0	**0**

For the original array the pattern indicated by the emboldened numbers yields the minimum allocation of

$$10 + 10 + 12 + 12 + 14 = 58.$$

Having negative numbers in the array does not affect the method because the stage of 'subtracting the least number in a row from each element of that row' will make all elements at least zero. For example, in Table 1.7 each element has -3 subtracted from it, which is just the same as adding 3 to each element.

8	-3	4	7	$-(-3)$ or $+3$ $\rightarrow$	11	0	7	10

Table 1.7

To maximise an allocation you can therefore simply change the sign of every element and then minimise in the usual way, as in Table 1.8.

3	1	2	Change sign $\rightarrow$	-3	-1	-2	$-(-3) \rightarrow$	0	2	1

Table 1.8

The same effect is produced by subtracting each element from the largest value in the array, shown in Table 1.9.

3	1	2	Subtract from 3 $\rightarrow$	0	2	1

Table 1.9

The next example illustrates this process.

Example 1.1.2
Choose five numbers from the array given below so that the sum of the five numbers is the greatest possible. No two numbers can be in the same row or column.

10	10	9	8	10
10	12	12	9	13
16	16	14	12	15
14	15	12	12	16
15	16	14	13	14

First, convert the problem to a standard minimising one by changing the sign of each element of the array or by subtracting each element from 16. Reducing by rows then gives:

0	0	1	2	0
3	1	1	4	0
0	0	2	4	1
2	1	4	4	0
1	0	2	3	2

Reducing by columns gives:

0	0	0	**0**	0
3	1	**0**	2	0
0	0	1	2	1
2	1	3	2	**0**
1	**0**	1	1	2

The numbers in bold type show the minimum allocation. For the original array the same pattern yields the maximum allocation of $16 + 16 + 12 + 8 + 16 = 68$.

1.2 The Hungarian algorithm

The opening example of Section 1.1 led to the reduced array in Table 1.10.

2	0	1	–	
0	0	0	0	
0	–	0	0	
1	0	1	3	

Table 1.10

At this stage you had to 'spot' that an allocation of four zeros was impossible and that it was necessary to use a 1. In larger and more complicated examples it is better to have a systematic procedure. One such method is called the **Hungarian algorithm**. To apply this algorithm start with an array that has been reduced by rows and columns. Cover all the zero elements with the minimum number of vertical or horizontal lines or both. For the array in Table 1.10 three lines are needed, as shown in Table 1.11.

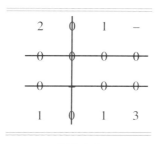

Table 1.11

Note the least uncovered element, in this case 1. Add this element to the elements of each covered row and then to the elements of each covered column. Where an element is covered twice, add the least uncovered element twice. See Table 1.12 for the result of this process.

2	1	1	–
1	2	1	1
1	–	1	1
1	1	1	3

Table 1.12

Then subtract this element from every element of the array. The result is shown in Table 1.13.

1	0	0	–		1	**0**	0	–
0	1	0	0		**0**	1	0	0
0	–	0	0		0	–	0	**0**
0	0	0	2		0	0	**0**	2

Table 1.13   Table 1.14

It is now possible to allocate four zeros in several ways; one is shown by the numbers in bold type in Table 1.14. This pattern, applied to the original array, yields the minimum value of $9+5+9+7=30$, that is £30,000.

The reason that this procedure further reduces the array is that the minimum number is added on $3\times 4=12$ times but is then subtracted 16 times. The following statement of the Hungarian algorithm shortens this procedure.

> **Hungarian algorithm**
>
> **Step 1** Reduce the array of costs by both row and column subtractions.
>
> **Step 2** Cover the zero elements with the minimum number of lines. If this minimum number is the same as the size of the array (which for a square matrix means the number of rows) then go to Step 4.
>
> **Step 3** Let m be the minimum uncovered element. The array is augmented by reducing all uncovered elements by m and increasing all elements covered by two lines by m. Return to Step 2. This process is called **augmenting the elements**.
>
> **Step 4** There is a maximal matching using only zeros. Apply this pattern to the original array.

Step 3 of the Hungarian algorithm is equivalent to the procedure used on the previous page: adding the least uncovered element to all elements covered by one line, adding twice the least uncovered element to all elements covered by two lines and then subtracting the least uncovered element from every element in the array. The longer procedure was presented first because it shows more clearly that the best matching remains the same at each stage.

Example 1.2.1

A company has four sales representatives to allocate to four groups of retailers. The table shows the estimated weekly mileage of each representative when assigned a particular group. How should the groups be allocated to minimise the total mileage?

	1	2	3	4
Alex	280	280	260	210
Ben	470	480	460	420
Charles	370	390	380	330
Davinia	220	250	240	220

The following array shows the situation after Step 1, after row and column reductions.

7	4	3	0
5	3	2	0
4	3	3	0
0	0	0	0

Numbers are tens of miles.

The next array, on the left, shows the situation after Step 2, where two lines are needed to cover the zeros. The right array shows Step 3, where 2, the minimum uncovered element, has been subtracted from all uncovered elements, and elements covered by two lines have been increased by 2. After this, the algorithm returns to Step 2.

CHAPTER 1: ALLOCATION

7	4	3	0		5	2	1	0
5	3	2	0		3	1	0	0
4	3	3	0		2	1	1	0
0	0	0	0		0	0	0	2

Steps 2 and 3 are carried out again, as before.

5	2	1	0		4	1	0	0
3	1	0	0		3	1	0	1
2	1	1	0		1	0	0	0
0	0	0	2		0	0	0	3

At this stage, carrying out Step 2 requires four lines. As this is the same as the size of the array, you can go to Step 4, where the pattern of zeros is shown in bold type.

4	1	0	0		4	1	**0**	**0**
3	1	0	1		3	1	**0**	1
1	0	0	0		1	**0**	0	0
0	0	0	3		**0**	0	0	3

Using the pattern of zeros, the optimum allocation is 1-Davinia, 2-Charles, 3-Ben, 4-Alex, with mileage $220 + 390 + 460 + 210 = 1280$.

1.3 Non-square arrays

Suppose five workers are available for four tasks. The times each worker would take at each task are in Table 1.15. How can each task be allotted to a different worker to minimise the total time?

	1	2	3	4
Angel	170	220	190	200
Britney	140	230	150	160
Cleo	180	210	170	170
Dimitri	190	240	210	220
Ed	160	220	170	170

Times are in minutes.

Table 1.15

To be able to apply the methods of this chapter it is first necessary to create a square array by adding a dummy column.

The trick is to add in a column of equal numbers so that this column does not influence the choice of workers for the other tasks. It is conventional (but not necessary) to make these numbers equal to the largest number in the array. This technique is shown in the next example.

Example 1.3.1
Select four numbers from the array given below so that the sum of the four numbers is the least possible. No two numbers can be in the same row or column.

170	220	190	200
140	230	150	160
180	210	170	170
190	240	210	220
160	220	170	170

Add in a column of 240s to obtain a square array.

170	220	190	200	240
140	230	150	160	240
180	210	170	170	240
190	240	210	220	240
160	220	170	170	240

Reducing rows and then columns, and dividing by 10, so that the entries are in tens, gives the following array, which requires three lines to cover the zeros.

0	1	2	3	2
0	5	1	2	5
1	0	0	0	2
0	1	2	3	0
0	2	1	1	3

After covering the first column, and the third and fourth rows, and augmenting the elements (only one return to Step 2 is required), you obtain the array on the next page in which the zeros in each row and column are in bold type.

0	**0**	1	2	1
0	4	**0**	1	4
2	0	0	**0**	2
1	1	2	3	**0**
0	1	0	0	2

The solution to the original problem is $160 + 220 + 150 + 170 = 700$.

To apply the Hungarian algorithm to a non-square array, first add in dummy rows or columns to make the numbers of rows and columns equal.

Exercise 1

1. The four members of a swimming relay team must, between them, swim 100 metres of each of backstroke, breaststroke, butterfly and crawl.

   Five hopefuls for the team have personal best times as follows.

	Back	Breast	Butterfly	Crawl
A	66	68	71	60
B	69	69	72	60
C	68	70	73	61
D	65	66	71	63
E	63	65	74	60

   Times in seconds for 100 metres.

   (a) Convert the array into a square array to which the Hungarian algorithm can be applied.

   (b) Hence find which four swimmers should be chosen and the stroke for which each should be used.

2. The scores of the four members of a quiz team on practice questions are as follows.

	Sport	Music	Literature	Science
Ali	16	18	17	14
Bea	19	17	14	18
Chris	12	16	16	15
Deepan	11	15	17	14

   A different person needs to be picked for each of the four different topics.

   (a) How could the table be altered in order to use the Hungarian algorithm?

   (b) Hence allocate the topics to the members of the team.

3 Apply the Hungarian algorithm to find the minimum possible total of six numbers, chosen from the table below in such a way that no two numbers lie in the same row or column.

3	2	1	3	1	2
2	1	3	1	2	3
3	4	5	2	5	3
4	3	2	1	3	4
5	4	3	3	2	3
3	1	4	1	2	1

4 Find the maximum possible total for six numbers chosen as in Question 3.

5 Suppose that the Hungarian algorithm is being applied to an array.

  (a) Suppose further that the zeros in a 5×5 array are covered by three lines and the least non-covered element is a 2. When the array is augmented once, what is the reduction in the total of all the elements?

  (b) Suppose that the zeros of an $n \times n$ array are covered by k lines and that the least non-covered element is l. What is the reduction in the total of all elements when this array is augmented once?

6 A builder has four labourers who must be assigned to four tasks. The estimates of the times each labourer would take for the different tasks are as shown in the table.

	Task			
	1	2	3	4
A	3	4	4	3
B	3	3	1	2
C	4	3	2	4
D	4	1	2	3

Times in hours.

Given that no labourer can be assigned to more than one task, use the Hungarian algorithm to find the optimum assignment.

7 A relay team has four runners who must be assigned to the four legs of a 4×2 mile race. The estimates of the times that each runner would take for the different legs are in the table.

	Leg			
	1	2	3	4
Haile	710	710	710	700
Josh	700	670	680	690
Stephen	670	660	670	680
Tom	680	670	660	710

Find the optimum assignment of runners.

Miscellaneous exercise 1

1. Four workers are to be allocated to four jobs. The cost, in £ thousands, of using each worker for each job is shown in the table.

		Job			
		Building	Carpentry	Drainage	Electrics
Worker	Jenny	4	3	2	7
	Kenny	3	2	3	4
	Lenny	6	3	4	5
	Penny	7	7	7	6

   (a) Use the Hungarian algorithm to pair the four workers with the four jobs to minimise the total cost.

   (b) Give the minimum total cost. (OCR)

2. Donna Wyg runs a detective agency. She has been employed to follow a suspect and needs to choose a different disguise for each day. Donna estimates that the job will last four days. She can choose from five disguises.

   Past experience has shown that some disguises are more appropriate than others for certain locations. She knows the likely locations of the suspect on each day.

   The table shows the probability that Donna will be spotted in each of the disguises at each of the locations.

Day	Location	Disguise				
		Artist	Bag lady	Cleaner	Duchess	Expert
Wednesday	Waxworks	0.12	0.24	0.12	0.44	0.18
Thursday	Theatre	0.14	0.27	0.10	0.46	0.43
Friday	Fair	0.16	0.23	0.13	0.45	0.51
Saturday	Station	0.14	0.22	0.11	0.47	0.12

   Donna wants to choose the disguises so that the sum of the probabilities of being spotted is as small as possible. To do this she will use the Hungarian algorithm.

   (a) The Hungarian algorithm requires the number of rows and columns to be equal. Add an appropriate dummy row to the probability matrix to make it square.

   (b) By reducing columns first, find a reduced probability matrix and explain how you know whether it gives a minimum probability allocation.

   (c) Write down a list showing which disguise Donna should choose for each day. (OCR)

3 Four house building companies, 1, 2, 3 and 4, are invited to tender for the construction of four different types of house L, M, S, T. Each company estimates the price it would charge. This information is shown in the table in £1000's.

		House type			
		L	M	S	T
Company	1	40	35	24	21
	2	38	44	28	16
	3	44	42	26	22
	4	43	39	40	20

Regulations prevent a building company from constructing more than one type of house.

(a) Use the Hungarian algorithm, by applying column reductions first, to determine which company should build which house if the total cost is to be a minimum.

(b) Calculate the minimum cost. (OCR)

4 The headmaster of a primary school is well-known for the strange combinations of shirt colours and tie patterns that he chooses. One day he brings four shirts and four ties to school and asks the children, by voting, to help him pair the ties with the shirts.

The number of votes cast against each combination is shown in the table.

		Tie pattern			
		Circles	Diamonds	Flowers	Gaudy
Shirt	Purple	5	7	12	26
	Orange	8	3	18	21
	Red	3	10	15	22
	Yellow	5	6	14	25

Use an appropriate algorithm to pair the four shirts with the four tie patterns so as to minimise the total number of votes against. (OCR)

5 Simon is planning a four-day break. He wants to spend one day at the art gallery, one day at the beach, one day at the castle and one day at the exhibition.

The cost, in £, for Simon to visit each of these places on each day is shown in the table.

	Art gallery	Beach	Castle	Exhibition
Wednesday	5.00	1.50	4.50	6.30
Thursday	4.50	1.20	5.00	6.00
Friday	5.00	1.50	5.00	6.00
Saturday	6.00	1.50	5.50	7.00

(a) Use the Hungarian algorithm to pair the four places with the four days so as to minimise Simon's total cost.

(b) Give the minimum total cost. (OCR)

6 Granny is on holiday and wants to send postcards to her five grandchildren. She has chosen six postcards and has given each card a score to show how suitable it is for each grandchild. A high score means that the card is more suitable.

		Card					
		Maps	Puppies	Railway	Seaside	Teddies	Views
Grandchild	Arnie	5	0	4	1	0	3
	Beth	3	1	0	1	0	3
	Cyril	0	3	2	3	2	1
	Des	3	2	2	3	2	2
	Erin	1	2	3	3	1	2

(a) The Hungarian algorithm finds the allocation with minimum total cost. Show how Granny's problem can be converted into a minimisation problem.

(b) The Hungarian algorithm requires the matrix to be square. Explain how to represent Granny's problem as a square matrix.

(c) Use the Hungarian algorithm, reducing columns first, to pair the cards to the grandchildren in the most appropriate way. (OCR)

7 Four coaches are to be allocated, one each, to train four rowing crews. They have an initial training session with each crew who are then timed over a fixed course with the following results.

		Crew			
		P	Q	R	S
Coach	Bert	85	88	93	84
	Carl	88	87	88	84
	Debbie	82	83	81	81
	Elaine	84	98	85	81

Use the Hungarian algorithm to suggest how to assign one coach to each crew by choosing the allocation which would minimise the total time achieved by this assignment in the initial training session. (AQA)

8 It is required to choose one element from each row and column of the table given below, in such a way as to minimise the total of the numbers.

1	2	0	0	1	2	1	0
1	0	1	0	2	0	1	1
0	3	2	0	1	2	2	0
1	0	1	2	0	0	1	1
2	3	0	2	2	1	3	1
3	2	0	0	1	1	1	0
1	1	0	0	0	2	0	1
1	2	0	1	2	2	2	1

(a) Find 7 lines covering all the zeros.

(b) Hence use the Hungarian algorithm to find the minimum total.

(c)* How many choices of numbers give the minimum total?

2 Network flows

This chapter is about maximising the flow through a network. When you have completed it you should

- know how to represent flow problems by networks with directed edges
- understand and know how to apply the maximum flow/minimum cut theorem
- know how to find a maximum flow in a network, subject to given constraints (including both upper and lower capacities), by using a labelling procedure to augment a given flow
- be able to deal with multiple sources and sinks, and with vertices with restricted capacities.

2.1 Some important terms

Problems involving flows through networks are widespread. They include the movement of traffic through an airport, the flow of oil in a system of pipes and the transfer of information across the internet.

Consider the problem of finding the maximum number of cars that can enter and leave the road network shown in Fig. 2.1.

The weights on the edges are called **capacities**. The start of the network is called the **source**, and is often labelled S. The end of the network is called the **sink**, and is often labelled T.

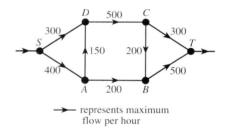

Fig. 2.1

For the problem above, you may have spotted a flow such as that in Fig. 2.2 which gives a total throughput of 650 cars per hour. In this solution, edges which are carrying their full capacity, such as SD and CB, are said to be **saturated**.

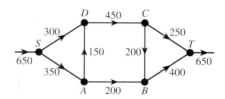

Fig. 2.2

There is a simple way of seeing that this flow is the maximum possible by making a suitable **cut** across the entire network.

A **cut** is any continuous line which separates S from T. A cut must not pass through any vertices.

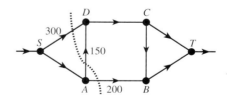

Fig. 2.3

In Fig. 2.3 the maximum possible flow across the cut line in the direction from S to T (called the **value**, or **capacity**, of the cut) is $300 + 150 + 200 = 650$ so this flow cannot be bettered unless the capacities of at least one of the roads SD, AD or AB is increased.

You should note that, throughout this chapter, it will be assumed that there can be no 'build-up' at a vertex. That is, with the exception of the source and sink, the sum of the flows into a vertex must be equal to the sum of the flows out of that vertex.

2.2 Maximum flow/minimum cut theorem

Any cut across a network can be thought of as a collection of potential bottlenecks which restrict the flow through the network.

Fig. 2.4 is an example of a cut, with five edges crossing it. Four of them, a, b, d and e, have flows from the source to the sink, but one of them, c, is in the reverse direction.

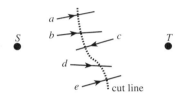

Fig. 2.4

The value of any cut limits the maximum possible flow. In this case the maximum flow cannot exceed $a + b + d + e$. Note that c is not included in the value of the cut because it is a flow in the wrong direction.

You have therefore seen that

> the value of any permissible flow $\leq$ the value of any particular cut.

In particular

> the maximum flow $\leq$ the value of the minimum cut.

This result should remind you of the results of DI Section 5.3, where equality actually held. The same is true here. It can be proved that the maximum flow equals the value of the minimum cut.

> **Maximum flow/minimum cut theorem**
>
> (a) The flow through a network cannot exceed the value of any cut.
>
> (b) The maximum flow equals the value of the minimum cut.

This theorem does not actually provide you with an algorithm to find the maximum flow. It does, however, give you a useful tactic when considering relatively simple networks.

> If you can find, by inspection, a flow and a cut which have the same value, then the maximum flow/minimum cut theorem tells you that you have obtained the maximum flow.

Example 2.2.1

(a) Find the value of the cut shown in the diagram.
(b) Find a flow with the same capacity as the value found in part (a). What can you deduce?

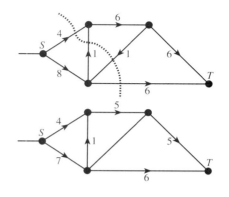

(a) The value of the cut is $4+1+6=11$.

(b) The diagram on the right shows a flow of 11. This is the maximum possible flow.

Exercise 2A

1 Find the values of the four cuts shown in the diagram. What can you deduce about the maximum flow?

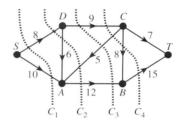

2 (a) Find the maximum flow through this network.

(b) Prove your answer cannot be exceeded by finding a cut of the same value.

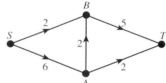

3 The numbers on the edges in the diagram represent the maximum numbers of passengers, in 1000s per day, who can be carried between five airports.

(a) What is the maximum number of passengers per day who can get to T from S?

(b) If the capacity of only one connection can be increased, which one should it be?

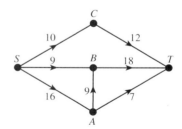

4 (a) Find the cut of minimum value for the network shown in the diagram.

(b) How do you know your answer to part (a) is the minimum?

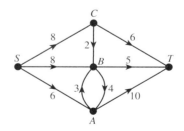

5 For the network shown, find a number k such that there is both a flow of k from S to T and a cut of value k.

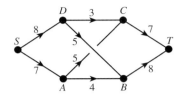

2.3 The labelling procedure

So far you have no algorithm to find the maximum flow in a large and complicated network. This section introduces an algorithm which can be used to systematically augment, that is improve, any initial flow. Note that there is no problem about finding an initial flow, because you can start with flows of zero!

To illustrate the procedure, first consider a very simple example. Fig. 2.5 shows a network with the maximum capacity shown for each edge.

Start with an initial flow of 5 along S-A-B-T. The next step is to redraw the network, indicating the amounts by which the flow along each edge could be altered.

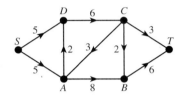

Fig. 2.5

For example, AB has a flow of 5 and a capacity of 8, so the flow from A to B could be increased by 3 from 5 to 8, or decreased by 5 from 5 to 0. AB is therefore redrawn as shown in Fig. 2.6. Similarly, AD has a flow of 0 and a capacity of 2, so it is redrawn as shown in Fig. 2.7.

You should then obtain the new network shown in Fig. 2.8, which is the **network of possible increases** (excess capacities, or 'how much more', and potential backflows, or 'how much less') associated with the original network and initial flow.

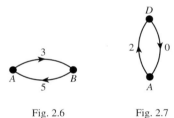

Fig. 2.6   Fig. 2.7

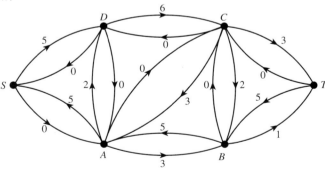

Fig. 2.8

You then find *any* flow from S to T on this new network. For example, you can see that a flow of 3 on S-D-C-T is possible. Adjust the labellings as shown in Fig. 2.9.

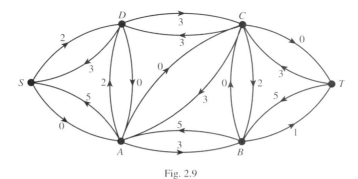

Fig. 2.9

You can now find another flow, 1 on *S-D-C-B-T*. Adjust the labellings again, as in Fig. 2.10.

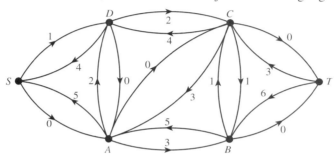

Fig. 2.10

If you look at the edges into *T* which show how much the capacity can be increased, you will see that they are both labelled 0. The original edges into *T* have become saturated. In the network of increases, *S* and *T* are now disconnected, so there can be no extra flow from *S* to *T*.

Therefore the maximum flow is the sum of 5 on *S-A-B-T*, 3 on *S-D-C-T* and 1 on *S-D-C-B-T*.

Here is the full algorithm.

Labelling procedure

Step 1 Begin with any initial flow or zero flow.

Step 2 Replace each edge with two edges, one showing the amount by which the flow can be increased (its excess capacity) and the other showing the amount by which the flow can be decreased (its potential backflow).

Step 3 If *S* is still connected to *T* (this can be determined by Dijkstra's algorithm if it is not obvious), then find a new flow from *S* to *T* and alter the excess capacities and potential backflows as necessary.

Step 4 Repeat Step 3 until *S* is disconnected from *T*.

If *S* is disconnected from *T* then the maximum flow is the sum of all the flows of Steps 1 and 3.

Note that, at the end of the labelling procedure, the maximum flow can also be found by subtracting the excess capacities leaving the source from the original capacities leaving the source, or by doing the same with the sink.

Thus the original capacity leaving the source was 10, and the excess capacity is now 1. This leaves the maximum flow as 9. You can check for yourself that the same is true for the sink.

Example 2.3.1
The first diagram shows the maximum capacities of edges in a network and the second diagram shows the flows currently established.

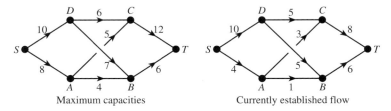

(a) Find a minimum cut and give its value.
(b) Explain how you know that the established flows are not maximum.
(c) Apply the labelling procedure to augment the flows and find a set of maximum flows.

    (a) The cut $\{DC, AC, BT\}$ has a value of 17.

    (b) The maximum flow must be 17 whereas the established flow is only 14.

    (c)

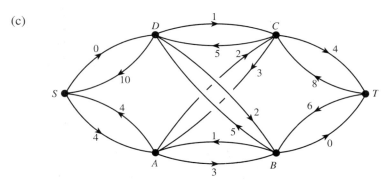

    Augment the flow by 2 on S-A-C-T and by another 1 on S-A-B-D-C-T.

The augmented flow can be in the opposite direction to the currently established flow, as in the case of the flow of 1 from B to D.

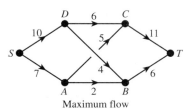

    The flow now has value 17 and therefore must be maximum. The final maximum flows are shown in the figure on the right.

2.4 Extensions

In practical applications, the methods developed in this chapter often need to be extended to cope with variations such as multiple sources or multiple sinks or both; vertices of restricted capacity; and edges with lower as well as upper capacities. For example, an oil pipeline system may have multiple inflows and multiple outflows; high-capacity roads may pass through a village where road works restrict capacity; a conduit may need a minimum level of use in order to remain free-flowing. This section and the next one will illustrate how you can tackle such problems.

Consider the network shown in Fig. 2.11. It has two sources, S_1 and S_2, and three sinks, T_1, T_2 and T_3.

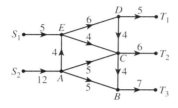

Fig. 2.11

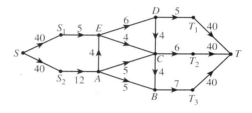

Fig. 2.12

In order to apply the methods of this chapter create a new **supersource**, S, that feeds the existing sources with capacities sufficiently big so as not to affect the solution. Similarly, create a new **supersink**, T. The new network obtained is shown in Fig. 2.12.

You can now apply the standard methods to the new network. The cut through S_1E and S_2A has value 17. Any flow of this value will therefore be a maximum flow. You can then obtain a maximum flow on the original network from this flow by removing the supersource and supersink, as shown in Fig. 2.13.

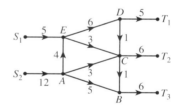

Fig. 2.13

If a network has more than one source, then create a new supersource.

If a network has more than one sink, then create a new supersink.

To this point, flows in a network have only been constrained by the capacities of the edges in the network. However, it is also common for flows to be restricted by the capacities of vertices. Suppose a network models foot traffic through a building where the edges represent corridors and the vertices represent doorways. A realistic analysis of flows through this network may have to consider that some of the doors are security or fire doors and, as a result, restrict the number of people who can pass through the building in a given amount of time.

Example 2.4.1

In the network shown in the diagram, vertex B has a restricted capacity of 8. Find the maximum possible flow.

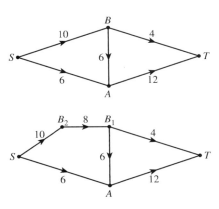

Replace vertex B by a pair of vertices B_1 and B_2, linked by an edge of capacity 8.

The minimum cut for the new network is 14, so this is also the maximum flow value.

> If a vertex has a restricted capacity, then replace it by two unrestricted vertices connected by an edge of the relevant capacity.

2.5 Minimum capacities

Sometimes, a minimum flow is essential along one or more of the edges. In such cases it is conventional to label each directed edge with two numbers, an upper capacity and a lower capacity. An example is shown in Fig. 2.14.

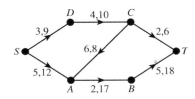

Fig. 2.14

For such a network, the value of a cut is defined as follows.

> The value of a cut is the sum of the upper capacities for edges that cross the cut line in the direction from S to T minus the sum of the lower capacities for edges that cross the cut line in the direction from T to S.

With this definition of the value of a cut, the maximum flow/minimum cut theorem can be applied to networks with both minimum and maximum capacities.

Example 2.5.1

(a) Find the values of the three cuts shown in the network.
(b) Find the maximum flow and the minimum cut.

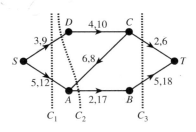

(a) Using the definition in the shaded box, the values of the cuts C_1 and C_3 are $12 + 9 = 21$ and $18 + 6 = 24$ respectively.

The cut C_2 has an edge AC flowing from T to S with a minimum capacity of 6 which must be subtracted. This means that the value of C_2 is $9 - 6 + 17 = 20$.

(b) A flow of 20 is given by   S-D-C   with capacity   9
                                S-A   with capacity   11
                                C-A   with capacity   6
                                A-B-T   with capacity   17
                                C-T   with capacity   3.

Because there is a cut with this value this is the maximum flow, and C_2 is the minimum cut.

The labelling procedure of Section 2.3 can be used without alteration with minimum as well as maximum capacities, except that you must now start with some initial feasible flow rather than zero flow.

Example 2.5.2
Apply the labelling procedure to the network of Example 2.5.1, shown in Fig. 2.15, to augment the initial flow given in Fig. 2.16.

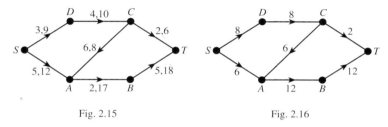

Fig. 2.15   Fig. 2.16

The network of increases associated with this example is shown in Fig. 2.17. This time the backflows are calculated using the minimum capacities. Thus, looking at the edge SD, where there is a minimum capacity of 3, the greatest amount by which the flow of 8 can be reduced is 5.

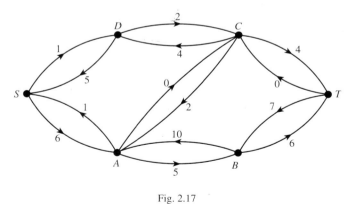

Fig. 2.17

The initial flow can be augmented by a flow of 5 on S-A-B-T and by a flow of 1 on S-D-C-T.

You can check that this labelling procedure gives the solution obtained in Example 2.5.1(b).

Exercise 2B

1. The diagrams below represent capacities and an initial flow. Draw a diagram showing excess capacities and potential backflows.

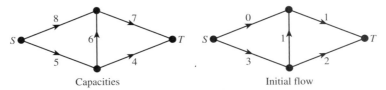

Capacities   Initial flow

2. The diagrams below represent lower and upper capacities, and an initial flow. Draw a diagram showing excess capacities and potential backflows.

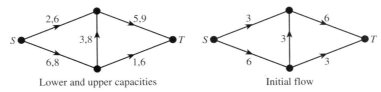

Lower and upper capacities   Initial flow

3. In this network of flows, the edges have unlimited capacities but the vertices have lower and upper capacities as shown.

    (a) Redraw the network in a form to which you could apply the maximum flow/minimum cut theorem.

    (b) Find both a maximum flow and a minimum cut.

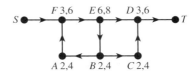

4. Redraw the network shown with two extra vertices: a supersource and a supersink.

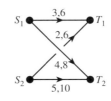

5. In the network shown, S represents an oil field, T a refinery and the other vertices are intermediate stations. Each edge depicts a pipeline through which oil can travel, the capacities being in millions of barrels per hour. Use the labelling procedure to find the maximum number of barrels per hour which can be moved to the refinery.

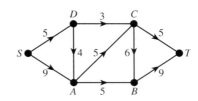

6 The diagram shows a system of pipes with lower and upper capacities.

   (a) Find a flow of 11 from S to T.
   (b) Use the labelling procedure to augment the flow of part (a). Hence find the maximum flow from S to T.
   (c) Find a cut of value equal to the maximum flow.

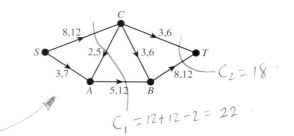

7 EZ-Fly has had to cancel its direct flight from Glasgow to Luton. The number of seats available on other EZ-Fly routes are as shown.

From	To	Number of seats
Glasgow	Liverpool	12
Glasgow	Manchester	6
Liverpool	Manchester	4
Liverpool	Luton	6
Manchester	Luton	8

Formulate and solve a maximum flow problem to determine how many passengers can be re-routed from Glasgow to Luton on EZ-Fly flights.

8* (a) Find the range of possible values of the flow in edge DB in this network of upper and lower capacities.
    (b) Hence find the ranges of possible values of x and y.

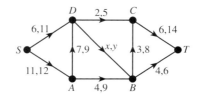

9 The daily capacities, in 1000s of people, of a national network of coaches are shown in the table.

		To				
		London	Birmingham	Manchester	Leeds	Edinburgh
	London	–	30	–	20	–
	Birmingham	–	–	20	10	–
From	Manchester	–	–	–	10	15
	Leeds	–	–	–	–	25
	Edinburgh	–	–	–	–	–

   (a) What is the maximum number of people per day who can use this company to travel from London to Edinburgh?
   (b) Find a minimum cut for the underlying network.

Miscellaneous exercise 2

1. The diagram represents a system of pipes. The weights show the (directed) maximum capacity for each pipe in litres per minute.

   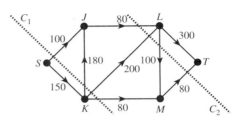

   (a) Calculate the values of the cuts C_1 and C_2 marked in the diagram.

   (b) Explain what the values calculated in part (a) tell you about the maximum flow from S to T. (OCR)

2. The diagrams show the maximum capacities of the edges in a directed distribution network, and the flows currently established in the network.

   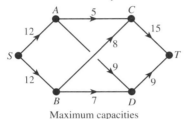
   Maximum capacities

   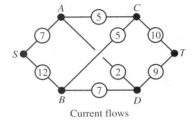
   Current flows

   (a) Find a minimum cut and give its capacity.

   (b) Say why the established flows do not give a maximum total flow through the network.

   (c) Use the labelling procedure to find a set of flows which do produce a maximum flow through the network. (AQA)

3. The figure shows a system of pipes with the lower and upper capacities, in litres per second, for each pipe.

   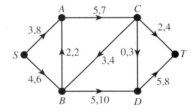

   (a) Draw a diagram showing a flow from S to T of 7 litres per second.

   (b) Augment your solution to part (a), using the labelling procedure and showing your working clearly, to find the maximum flow from S to T. (OCR)

4. The figure shows a system of pipes and the upper and lower capacities of each pipe, in litres per second.

   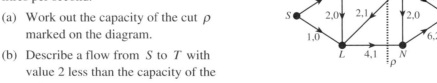

   (a) Work out the capacity of the cut ρ marked on the diagram.

   (b) Describe a flow from S to T with value 2 less than the capacity of the cut ρ.

   (c) What can you deduce from the results of parts (a) and (b)? (OCR)

5 The diagram shows a network of pathways in a maze. Rats move steadily through the maze.

Vertex S represents the entrance and vertex T represents the exit. The weights on the edges show the maximum number of rats per minute that can move along each pathway.

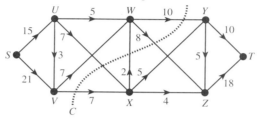

(a) Obtain the value of the cut marked C on the diagram.

(b) What can you deduce from your answer to part (a) about the flow of rats through the maze?

(c) Use a labelling procedure to construct a flow of 23 rats per minute.

(d) Write down the edges that are saturated in your answer to part (c). What can you deduce from this about the flow of rats through the maze. Explain your reasoning.

(OCR, adapted)

6 The linear programme below is to find a maximum flow from S to T in the network shown. The numbers show the maximum allowable flows along the edges.

Flows that are established along edges are indicated by x-values. For instance, the value of x_{SA} gives the flow along the edge SA.

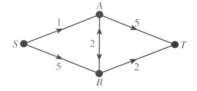

A computer package is used to solve the linear programme.

Maximise $\quad F = x_{SA} + x_{SB}$

subject to $\quad x_{SA} + x_{BA} - x_{AT} - x_{AB} = 0,$
$\qquad\qquad x_{SB} + x_{AB} - x_{BA} - x_{BT} = 0,$
$\qquad\qquad x_{SA} \leqslant 1, \quad x_{AT} \leqslant 5, \quad x_{AB} \leqslant 2,$
$\qquad\qquad x_{BA} \leqslant 2, \quad x_{SB} \leqslant 5, \quad x_{BT} \leqslant 2,$
$\qquad\qquad$ all x-values $\geqslant 0$.

The solution given by the computer package is

F	x_{SA}	x_{SB}	x_{AB}	x_{BA}	x_{AT}	x_{BT}
5	1	5	4	0	2	3	2

(a) What maximum flow is given by this solution? Give a cut to prove that this flow is a maximum flow. Say why your cut proves that the flow is maximal.

(b) The two equality constraints refer to vertices A and B respectively. Explain briefly the purpose of the equality constraints.

(c) Explain the structure of the objective function, and give an alternative objective function.

(AQA)

7 The figure shows a system of motorways and the maximum capacities, in thousands of vehicles per hour, on each motorway.

(a) Calculate the values of the cuts C_1 and C_2.

(b) Copy and complete the table below, showing every possible cut and its capacity.

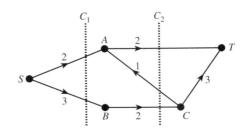

Edges in cut	Vertices on source side	Vertices on sink side	Capacity of cut
SA, SB	S	A, B, C, T	
SA, BC	S, B	A, C, T	

(c) State the maximum flow from S to T.

(OCR, adapted)

8 The figure shows a system of corridors and the maximum number of people who can move along the corridor (in either direction) each minute. When the fire bell rings all the people in the three rooms S_1, S_2 and S_3 must move along the corridors to one of the two fire assembly points T_1 or T_2.

(a) Show how the network can be modified so that it has a single source, S, and a single sink, T. (You need not redraw the whole network.)

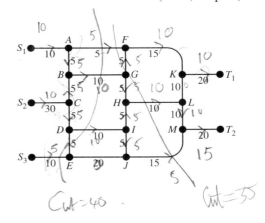

(b) A cut, γ, separates the vertices of the figure into the two sets X and Y, where $X = \{S_1, S_2, S_3, A, B, C, D\}$ and $Y = \{E, F, G, H, I, J, K, L, M, T_1, T_2\}$. Ignoring the rest of the network, calculate the maximum number of people per minute who can cross γ from X to Y.

(c) Give a flow in which exactly 25 people per minute move from S to T.

(d) (i) Find the maximum number of people per minute who can move from S to T.

(ii) Use the maximum flow/minimum cut theorem to show that the flow in (i) is maximal.

(OCR, adapted)

9 The figure shows a system of telephone cables and the maximum number of telephone lines, in thousands, that each cable can carry.

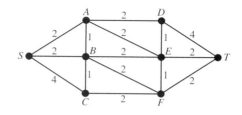

(a) Draw a diagram showing how S and T can be joined by five thousand lines.

(b) Augment your answer to part (a), using a labelling procedure and showing your working clearly, to find the maximum number of lines that the system can carry between S and T.

(c) Use the maximum flow/minimum cut theorem to verify that your answer found in part (b) is maximal. (OCR, adapted)

10 The figure shows a network of directed edges. The values on the edges show the maximum and minimum capacities, in litres per second.

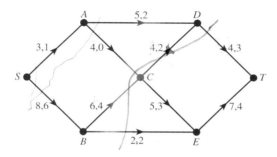

(a) The flow through vertex C must be at least 5 litres per second. Explain how to modify the network to show this, using directed edges with maximum and minimum capacities.

(b) A cut, γ, separates the vertices into the two sets $\{S, A, B, D\}$ and $\{C, E, T\}$. Calculate the maximum possible flow across γ.

(c) Explain what the value calculated in part (b) tells you about the maximum flow through the network. (OCR, adapted)

11 (a) Find the maximum value of the cut through edges AT, BC and SC in the network shown in the figure (minimum and maximum flows along edges are marked).

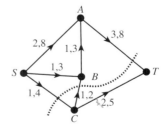

(b) Draw the network with a flow equal to the value of the cut found in part (a) which uses the maximum capacity of edge BA. Explain why this flow is a maximum.

(c) A new edge is added to the network from B to T with maximum flow of 3 and no minimum flow. By considering a flow augmentation (i.e. labelling procedure) of the flow in part (b), find the maximum flow of the new network and draw the network with the new solution. (AQA)

3 Critical path analysis

This chapter is about scheduling large projects. When you have completed it you should

- know how to construct and interpret activity networks
- be able to find earliest and latest start times by performing forward and reverse passes
- be able to identify critical activities and find a critical path
- know how to construct and interpret cascade diagrams and resource histograms
- understand how to carry out resource levelling.

3.1 Activity networks

Fig. 3.1 represents just one part of the process followed by the Sainsbury supermarket chain when launching a new product.

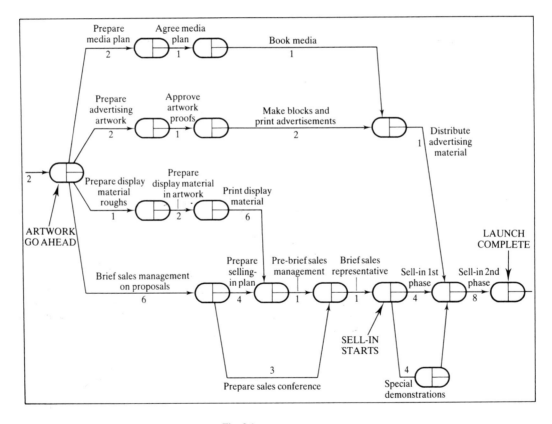

Fig. 3.1

This diagram is an example of an **activity network**. Such networks are useful because they express a project in a form which is visually meaningful, indicating, for example, which tasks must be done in sequence and which can be done at the same time. More importantly, activity networks allow questions of scheduling and resource use to be tackled systematically, using **critical path analysis.**

The techniques of critical path analysis were developed in the late 1950s by the US Navy and DuPont to plan and monitor such large-scale projects as the Polaris missile programme. Their general aim is to determine the minimum time in which a project can be completed and to optimise the use of manpower and resources in achieving this minimum completion time.

The starting point of critical path analysis is to break a project down into individual activities and to draw up a table showing the duration of each activity and which activities must be completed before a given activity can be started.

Consider a simple decorating project which involves the activities shown in Table 3.2.

	Activity	Duration (hours)	Immediately preceding activities
A	Prepare walls and ceiling	2	–
B	Sand woodwork	1	–
C	Prepare floor	1	–
D	Paint woodwork	2	A, B
E	Allow paint to dry	8	D
F	Emulsion walls and ceiling	3	E
G	Lay new flooring	2	C, F
H	Construct shelving unit	3	–
I	Fix unit	1	H, F

Table 3.2

When constructing an activity network, it is conventional to let the vertices of the network represent the activities. The vertices are drawn as boxes to leave space for numbers which will be explained in the next section.

The network should be drawn so that preceding activities are all shown to the left of the given activity. For example, D is preceded by A and B so you should draw A and B to the left of D, as in Fig. 3.3.

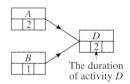

Fig. 3.3

The numbers in the middle-bottom boxes indicate the durations of the activities.

Some textbooks use the alternative convention of letting the edges represent the activities. Fig. 3.1 shows such a method.

The situation involving G and I is more complicated. Activities C and F precede G, whereas H and F precede I (Fig. 3.4).

The whole network representing the activities in Table 3.2 is shown in Fig. 3.5 on the next page. A final activity, 'END' of duration 0, has been added so that the network has a

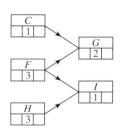

Fig. 3.4

single final activity. A single initial activity of duration zero could also have been added, but this is not necessary.

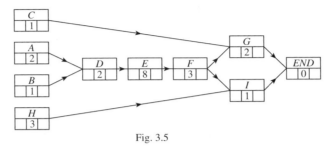

Fig. 3.5

Activity networks such as this are much easier to understand if you arrange them so that an activity which follows another is always to its right.

> **Procedure for drawing an activity network**
>
> **Step 1** On the left-hand side of the diagram draw a line of boxes to represent all those activities that have no immediate predecessors.
>
> **Step 2** To the right of any previous drawing, draw a line of boxes to represent all those activities for which all predecessors have already been drawn.
>
> **Step 3** Repeat Step 2, until all activities have been considered.
>
> **Step 4** Draw a box on the right-hand side of the diagram, to represent the finish.
>
> **Step 5** Draw directed edges to represent all immediate predecessors.

In practice, computers are used to draw activity networks, and to carry out the optimisation processes.

3.2 Earliest and latest starting times

An important aspect of critical path analysis is determining the possible starting and finishing times of activities. The earliest time after the start of a project that an activity can be started is called the **earliest start time**, or the **early time**.

It is useful to adopt a standard notation for labelling the vertices with times. The early time for an activity is placed in the box on bottom left, as in Fig. 3.6.

The early times are calculated by performing a **forward pass**, that is working through the activity network, from left to right, according to the early time algorithm.

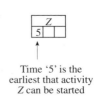

Time '5' is the earliest that activity Z can be started

Fig. 3.6

Early time algorithm (forward pass)

Step 1 Label all activities with no predecessors with early time 0.

Step 2 Choose any activity, X say, such that all activities to its left have already been given an early time.

Step 3 Consider all activities preceding X. Label X with an early time equal to the maximum value of

early time of preceding activity + duration of preceding activity

for all these activities.

Step 4 If all activities have been labelled then stop. Otherwise, return to Step 2.

Example 3.2.1
Find the early time for vertex X in the activity network shown in the diagram.

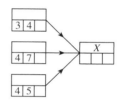

Three activities lead into X. The early time for X is therefore the maximum of

$3 + 4$, $4 + 7$, $4 + 5$.

The maximum of these is 11, so the early time is 11.

Note that the early time algorithm is really just common sense. How early you can start an activity depends on how soon the tasks on which it depends can be completed.

Example 3.2.2
When is the earliest that the project shown in the activity network on the right can be completed?

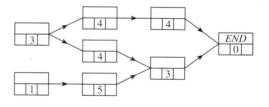

Applying the early time algorithm gives the early times shown below.

The earliest time the project can be completed is shown in the box on the far right, which is 11.

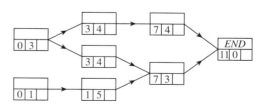

The **late time** for an activity is the latest that it can finish and still not delay the completion of the project.

Once you have determined all the early times you can calculate the late times by performing a **reverse pass**, that is working back through the network from right to left, according to the late time algorithm. The late times are written in the box on the bottom right of the vertex.

Late time algorithm (reverse pass)

Step 1 Label the final activity with a late time equal to its early time.

Step 2 Choose any activity, X say, such that all activities to its right have already been given a late time.

Step 3 Consider all activities immediately following X. Label X with a late time equal to the least value of

late time of following activity − duration of following activity

for all these activities.

Step 4 If all activities have been labelled then stop. Otherwise, return to Step 2.

Example 3.2.3

Find the late time for activity X.

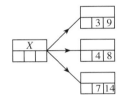

Three activities lead out of X. The late time for X is therefore the least of

$9 - 3$, $8 - 4$, $14 - 7$.

The least of these is 4, so the late time for activity X is 4.

Once again, the application of the algorithm should be common sense. How late you can finish an activity depends on how much time is available, and what else must be done, before the completion of the project.

Example 3.2.4

Complete the early and late times for Example 3.2.2.

Applying the late time algorithm to the solution of Example 3.2.2 gives the solution below.

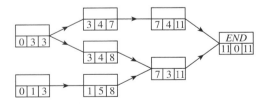

The early and late times for the activities refer to the earliest possible starts and the latest possible finishes. It is straightforward to use these to determine the latest start times and the earliest finish times for the activities.

Chapter 3: Critical Path Analysis

Example 3.2.5

One activity of a network is shown in the diagram. Determine the latest start and earliest finish times for activity A.

The earliest start time for activity A is 4. The earliest finish time is therefore $4 + 2 = 6$.

The latest finish time for A is 11. The latest start time is therefore $11 - 2 = 9$.

Exercise 3A

1 Draw activity networks for the projects with activities as given in the tables.

(a)

Activity	Duration	Immediately preceding activities
A	4	–
B	8	–
C	9	A
D	3	B
E	6	C
F	2	C, D
G	5	F

(b)

	Activity		Duration (min)	Immediately preceding activities
A	Pre-heat oven		10	–
B	Grease tin		0.5	–
C	Cream fat and sugar		2	–
D	Beat in eggs		0.5	C
E	Fold in flour		1	D
F	Put mixture in tin		0.5	B, E
G	Bake cake		10	A, F

2 Complete the early and late times for this activity network.

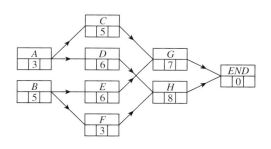

3 Complete the early and late times for the activity networks described in Question 1(a) and 1(b).

4 Find the earliest possible completion time for the project with activity network as shown.

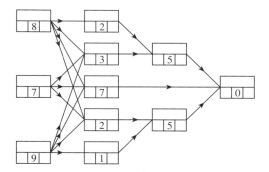

5 Four activities, their expected durations and precedences are shown in the table.

(a) Draw an activity network to represent these activities and their precedences.

(b) Determine the earliest and latest starting time for each activity, for completion of the project in the minimum time.

Activity	Duration	Immediately preceding activities
A	3	–
B	2	–
C	1	A, B
D	4	A, B

6 A construction project is divided into nine activities as shown.

(a) Construct an activity network for the project.

(b) Determine the earliest and latest starting times for each activity.

Activity	Duration (weeks)	Immediate predecessors
A	2	–
B	2	–
C	1	A
D	2	A, B
E	3	B
F	1	C
G	4	C, D, E
H	3	E
I	3	F, G, H

3.3 Critical activities

Consider again the activity network from Example 3.2.4, which is shown in Fig. 3.7.

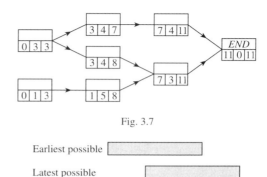

Fig. 3.7

The activity of duration 5 can start as early as time 1 and must finish by time 8. This information is illustrated in Fig. 3.8, where the shaded rectangles show the earliest and latest possible periods during which the activity takes place.

For this activity the two periods are not the same. There is therefore flexibility in the scheduling of this task. The technical term for this flexibility in scheduling is **float**.

Earliest possible

Latest possible

0 1 2 3 4 5 6 7 8

Fig. 3.8

However, other activities have no such flexibility. They are called critical activities because their prompt completion is critical to whether or not the project can be carried out in the least possible time.

Once you have determined the early and late times, you can easily pick out critical activities because

- their duration is the difference between the labels of their left-hand and right-hand boxes.

An example is the activity at the top right of Fig. 3.7, shown isolated in Fig. 3.9.

$\boxed{7\ |\ 4\ |\ 11}$ $11 - 7 = 4$

Fig. 3.9

> A **critical activity** is an activity for which there is no scheduling flexibility.

Since the successful completion of critical activities is essential to the prompt completion of the entire project, these are the activities which must be monitored most closely and to which manpower and machinery must be reallocated if there are any snags.

Each activity network has at least one path of critical activities leading from the start vertex to the final vertex. This path is called the **critical path**.

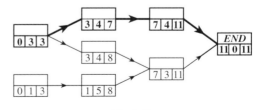

Fig. 3.10

An example of a critical path for the activity network from Example 3.2.4 is shown by the thick path in Fig. 3.10.

> A **critical path** is a path of critical activities leading from the start vertex to the final vertex.

Exercise 3B

1. What is the critical path for the network shown in the diagram?

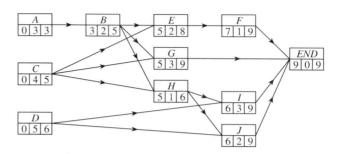

2. Suppose that for the network in Question 1 just one of activities *E*, *G*, *I* and *J* can be speeded up. Which one would you choose to accelerate? Explain your answer.

3. Complete the early and late times for the network shown when the duration of activity *X* is

   (a) $t = 4$;

   (b) $t = 2$.

   For what range of values of t is activity *X* critical?

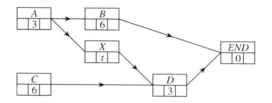

4. Henry and Lucinda are organising a children's party for their young daughter Abigail. The table lists the activities involved, their expected durations and precedences.

	Activity	Duration (days)	Preceded by
A	Decide who to invite	2	–
B	Find a suitable room and book it	5	–
C	Send out invitations and get replies	8	A, B
D	Book entertainment	3	B
E	Book caterer	2	C
F	Buy party bags	1	C
G	Buy Abigail a new party dress	3	D

(a) Draw an activity network to represent these activities and their precedences.

(b) Determine the earliest and latest starting time for each activity, for completion of the party arrangements in the minimum time.

(c) Identify the critical path and find the minimum time for completion of the party arrangements.

Lucinda and Henry find that the replies all come much sooner than they had expected, and so although they had allowed eight days for activity *C* it only takes three days.

(d) Identify the critical path assuming that activity *C* only takes three days. (OCR)

5   The table gives details of five tasks which have to be completed to finish a project.
    (a) Produce an activity network for the project.
    (b) Perform a forward pass and a backward pass to find the minimum time to completion and the critical tasks.

Activity	Duration	Immediate predecessors
A	5	–
B	2	–
C	1	B
D	2	A, B
E	3	A, C

6   (a) Draw an activity network for the project whose activities are described in the table.
    (b) Use an appropriate algorithm to find the critical path and the minimum completion time.
    (c) Which activity can have its start time delayed by the greatest possible time without affecting the overall completion time?

Activity	Duration (weeks)	Immediate predecessors
A	2	–
B	4	–
C	1	A, B
D	3	B
E	6	C
F	1	C, D
G	2	D

7   A project has the activity network shown. The times are given in days.
    (a) Determine the critical path and project duration.
    (b) By how much will the project be delayed if the duration of each activity is increased by two days?

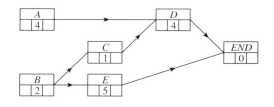

3.4 Cascade diagrams

Once the early and late times have been calculated, a cascade diagram is a good, easy to understand, way of displaying the information. Fig. 3.11 is an example of a cascade diagram.

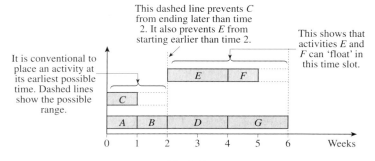

Fig. 3.11

You can easily spot the following features of this project.

- The project can be completed in a minimum of six weeks.
- Activities *A*, *B*, *D* and *G* cannot 'float', and are therefore critical.
- Activity *C* precedes *D*, and can start at any time during the first week.
- Activity *E* precedes *F*, and can start at any time during the third week.

When constructing a cascade diagram (also called a Gantt chart or diagram) it is a good idea to deal first with all the critical activities since their time slots are completely fixed.

Example 3.4.1
Construct a cascade diagram for a project with the activities given in the table.

Activity	Duration (weeks)	Immediate predecessors
A	5	–
B	3	–
C	2	–
D	3	A, B
E	4	B, C

Start by constructing an activity network with the early and late times. This is shown below.

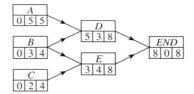

You can see that the critical path is *AD*, so start the cascade diagram by putting these across the bottom of the diagram.

Then note that the earliest start times of *B* and *C* are both zero, so this allows you to place them at the left of the diagram.

Finally, *E* can float between 3 and 8, and is placed at its earliest possible time, which is 3.

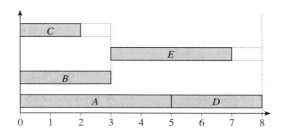

3.5 Resource levelling

So far, you have only been concerned with the time scale of activities and not what resources are needed. For example, the number of people needed for each task has not been considered. For most applications the availability and costs of resources are just as vital as the time scale. A first step in dealing with these issues is to draw a **resource histogram**.

Consider, for example, Fig. 3.12 which shows the cascade diagram for a project and notes the resources required for the various activities.

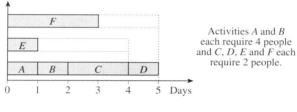

Activities A and B each require 4 people and C, D, E and F each require 2 people.

Fig. 3.12

The numbers of people required for each day are then shown in the resource histogram in Fig. 3.13.

For some businesses, it may be appropriate to use resources in this way, with more people needed at the beginning of the project. However, for many applications it is better to even out the use of resources as much as possible. For example, small construction firms will often employ a number of full-time staff. It is important that these staff are always fully employed rather than doing nothing. Most firms would want to avoid employing temporary labour if at all possible.

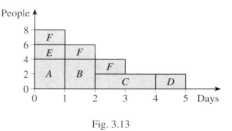

Fig. 3.13

The smoothing out of the usage of resources is called **resource levelling**. For the project shown in Fig. 3.13, resource levelling can be achieved as shown in Fig. 3.14, with the new resource histogram shown in Fig. 3.15.

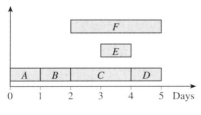

Fig. 3.14

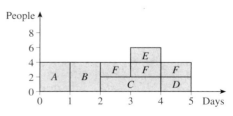

Fig. 3.15

Example 3.5.1

For the project with activities as shown, use resource levelling to find a schedule and an allocation of personnel which enables the project to be completed as soon as possible, but using at most three people at any one time.

Activity	Duration (days)	People needed	Immediate predecessors
A	2	2	–
B	4	1	–
C	1	2	–
D	3	1	C
E	1	1	A
F	1	1	B, D
G	3	2	B, E

Start by drawing the activity network and identifying the critical path, which is BG.

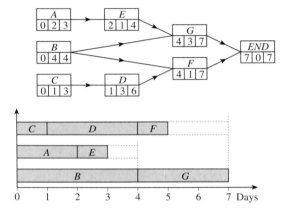

Then draw the cascade diagram, starting along the bottom with the critical path.

From the cascade diagram, you can see that activities A and C can be done in either order. A possible resource histogram is shown on the right.

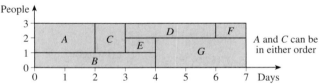

A and C can be in either order

Exercise 3C

1 The network represents the activities involved in a project and their durations in hours.

(a) Perform a critical path analysis to find the minimum project completion time and the critical activities.

(b) Construct a cascade diagram for the project, assuming activities start as early as possible.

(c) Assuming that each activity requires one person, state the maximum number of people required for the schedule of part (b). How can the project be completed in the minimum time but using at most two people at any one time?

2 The table shows the activities involved in a project, their durations, their immediate predecessors, and the number of people needed for each activity.

Activity	A	B	C	D	E	F
Duration (days)	1	2	1	1	4	2
Immediate predecessors	–	–	A, B	B	B	C, D
People needed	1	3	2	2	2	2

(a) Draw an activity network for this project.

(b) Perform forward and backward passes in order to find the critical path and the minimum project duration.

(c) Schedule the activities so that the project is completed as quickly as possible, using no more than four people at any one time. (AQA)

3 The activities involved in organising a charity event are shown in the activity network, where the durations are in days.

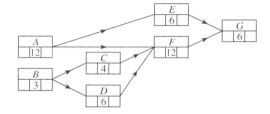

(a) Find the critical path and its duration. State the earliest and latest starting and finishing times for each activity.

(b) Given that each activity requires one helper in order to complete the project on time, find the minimum number of such helpers.

(c) The organiser has to cope with one fewer helper than required in part (b). Find the least possible delay to the completion of the project and state how this can be achieved.

4 A construction firm is involved in a building project subdivided into six major tasks as shown in the table.

Activity	Duration (days)	Workers needed	Immediate predecessors
A	14	2	–
B	6	1	–
C	4	3	B
D	10	3	A, C
E	20	1	B
F	16	4	B

(a) Assuming that enough workers are available, construct an activity network for the project. Find the critical path and the shortest possible finishing time for the building.

(b) Construct a resource histogram for the number of workers required, assuming that each task is to start as early as possible.

(c) Given that the firm employs eight full-time workers, describe how to schedule the activities so as to enable the earliest possible completion time without requiring the employment of additional workers.

Miscellaneous exercise 3

1 The table gives details of a set of six tasks which have to be completed to finish a project. The 'immediate predecessors' are those tasks which must be completed before a task may be started.

Task	Duration (days)	Immediate predecessors
A	2	–
B	8	–
C	5	A
D	4	A
E	2	A, B
F	2	B, C, D

(a) Produce an activity network for the project.

(b) Perform a forward pass and a backward pass to find the minimum time for completion and the critical tasks.

The following table shows, for each task, the extra cost that would be incurred in using extra resources to complete the task in one day less than the normal duration.

Task	A	B	C	D	E	F
Extra cost (£), for completion in 1 day less	1000	1250	750	500	1000	1000

(c) Find the minimum extra cost to complete the project in one day less than the minimum completion time which you found in part (b). (AQA)

2 The table lists six activities involved in painting a room.

	Activity	Duration (hours)	Preceded by
A	Prepare walls etc.	5	–
B	Undercoat walls	2	A
C	Paint ceiling	1	A
D	Paint walls	2	B, C
E	Paint woodwork	4	B, C
F	Tidy up	1	D, E

(a) Draw an activity network to represent these activities and their precedences.

(b) Determine the earliest and latest starting time for each activity.

(c) Identify the critical path, and the minimum time for completion of the painting, from start to finish, assuming that there is sufficient manpower available.

A start lag of 1 hour is to be incorporated between activities B and E.

(d) Describe what the start lag represents in relation to the activities B and E, and the effect that it will have on the critical path. (OCR)

3 A building project has been divided into a number of activities as shown in the table.

	Activity	Immediate predecessors	Duration (days)
A	Prepare site	–	1
B	Erect walls	A	4
C	Excavate and lay drains	A	7
D	Fit window frames	B	3
E	Erect roof	B	9
F	Install electrics	B	7
G	Install plumbing	C, D	6
H	Insulate roof	E	3
I	Plaster walls	F, G, H	2

(a) Construct an activity network for the project.

(b) Find the earliest start time for each activity for completion in the minimum time.

(c) Find the latest finish time for each activity for completion in the minimum time.

(d) Identify the critical path and state the shortest completion time for the whole project.

(e) Construct a cascade diagram for the project, assuming each activity is to start as early as possible.

(f) Each activity requires one worker. Draw a resource histogram showing the number of workers required each day.

(g) Given that there are only three workers available on any day, show how the activities could be allocated so that the project may be finished in the shortest completion time.

(AQA, adapted)

4 A major construction project is to be undertaken and the project has been divided into 11 activities, as shown in the table.

(a) Construct an activity network for the project.

(b) Find the earliest start time for each activity.

(c) Find the latest finish time for each activity.

(d) Identify the critical path and state the shortest completion time for the whole project.

Activity	Immediate predecessors	Duration (weeks)
A	–	3
B	–	4
C	A	2
D	B	3
E	C, D	5
F	C	6
G	D	7
H	E, F, G	4
I	G	6
J	F	8
K	H, I, J	3

Some of the activities can be speeded up at an additional cost. The following table lists the activities that can be speeded up, their additional cost in £ per week, together with the minimum times required to complete the activities.

Activity	Additional cost (£ per week)	Minimum time (weeks)
E	5000	1
I	4000	4
J	3000	5

The company wants to complete the project as soon as possible.

(e) (i) Find which activities should be speeded up. For each such activity, state, with justification, the reduction in the number of weeks.

(ii) Hence state the revised minimum time for the completion of the whole project.

(iii) Calculate the total additional cost the company would incur in meeting this revised completion time. (AQA)

5 The table at the top of the next page details the tasks, procedures and times associated with a project. The project is complete when all tasks are completed.

(a) Draw an activity network for this project.

(b) Determine the earliest start time and latest start time for the tasks, for completion in minimum time.

(c) Use your table from part (b) to identify the slack time available for each task. Hence, write down the critical path and the minimum time to complete the project. (OCR)

Task	Preceded by	Time to complete task (days)
A	–	7
B	–	5
C	A	6
D	A	4
E	B	5
F	B	7
G	E	4
H	F	6
I	C, D	9
J	G, H, I	3

6 The production of a leaflet to advertise a swimming pool involves 11 activities, as shown in the table.

	Activity	Duration (days)	Preceded by
A	Rough plan	3	–
B	Get finance	6	A
C	Detailed plan	5	A
D	Take photographs	4	B, C
E	Write text	6	B, C
F	Edit copy	2	D, E
G	Send to printer	1	F
H	Canvass likely interest	3	F
I	Check print proofs	1	G
J	Have leaflets printed	6	H, I
K	Send out leaflets	1	J

(a) Draw an activity network to illustrate these activities and their precedences.

(b) Determine the earliest and latest start times for each activity.

(c) Identify the critical path and the minimum time for completion for the production of the leaflets, from start to finish.

(d) The production team realise that they can begin to canvass likely interest (activity H) before editing copy (activity F) is completed. Explain briefly how this will alter the critical path and the minimum time for completion. (OCR)

7 The table lists six activities in a project, together with their durations, precedences and the number of people required for each activity.

Activity	Duration (days)	Preceded by	People required
A	3	–	2
B	2	–	3
C	1	A, B	2
D	5	B	4
E	4	C	2
F	5	D, E	1

In addition to the precedences shown in the table, activity *E* must not start until at least four days after activity *B* has started.

(a) Draw an activity network to represent these activities and their precedences.

(b) Determine the earliest and latest starting times for each activity, for completion of the project in the minimum time.

(c) Identify the critical activities, and the minimum time for completion of the project, assuming that there are sufficient people available.

(d) Find the minimum number of people required at any given time to complete the project in the minimum time. Explain your reasoning carefully. (OCR)

8 The table lists nine activities involved in building an attic extension, their durations and precedences.

	Activity	Duration (days)	Preceded by
A	Draw up plans and get planning permission	45	–
B	Clear rubbish and prepare site	16	–
C	Order stairs and window frames	12	A
D	Build new entrance and fit stairs	2	B, C
E	Install electrics	4	A, B
F	Fit floor	3	D, E
G	Fit window	1	D
H	Plaster walls	1	G, F
I	Decorate	3	H

(a) Draw an activity network to represent these activities and their precedences.

(b) Determine the earliest and latest starting times for each activity, for completion of the extension in the minimum possible time.

(c) Identify the critical path, and the minimum time for completion of the extension.

To save time, activity *C* (order stairs and window frames) is brought forward so that it happens at the same time as activity *A*.

(d) Work out how many days this will save on the whole project. (OCR)

9 The table lists activities which form a project, with their durations and precedences.

Activity	Duration (min)	Preceded by
A	5	–
B	2	A
C	4	A
D	5	B
E	4	B, C
F	3	C
G	2	D, E
H	3	E, F
I	4	G, H

(a) Draw an activity network to represent the project.

(b) Determine the earliest and latest starting time for each activity, for completion of the project in the minimum possible time.

(c) Identify the critical path, and the minimum time for completion of the project.   (OCR)

10 Paul is preparing a meal for his girlfriend. The table lists the activities involved, together with their durations and precedences.

	Activity	Duration (min)	Preceded by
A	Prepare table	5	–
B	Chop vegetables	8	–
C	Mix spices	5	–
D	Cook rice	20	–
E	Cook curry	10	B, C
F	Serve chutneys	2	A
G	Cook poppadoms	5	F
H	Put food on plates	2	D, E, G

(a) Draw an activity network to represent the preparation of the meal.

(b) Assuming that Paul can carry out two or more activities simultaneously, use a forward pass to determine the minimum time for preparing the meal.

(c) Use a backwards pass to identify the critical activities for preparing the meal in the minimum time, with the assumption from part (b).

Activities D, E and G may be carried out simultaneously, and activity D may be carried out at the same time as any activity except activity H. Otherwise Paul cannot carry out more than one activity at a time.

(d) Calculate the minimum time for the preparation of the meal under these conditions.
    (*You do not have to use a cascade diagram, but you may do so if you wish.*)   (OCR)

11 A group of workers are involved in a self-build project. The project has been divided into the activities shown in the table below.

Activity	Immediate predecessors	Duration (days)	Number of workers required
A	–	2	8
B	A	6	4
C	A	12	4
D	B	12	8
E	B	16	4
F	B	4	6
G	C, F	10	6
H	E	4	4
I	D, G, H	2	8

(a) Assuming that sufficient workers are available,

(i) construct an activity network for the project,

(ii) find the earliest start time for each activity,

(iii) find the latest finish time for each activity,

(iv) identify the critical path and state the shortest completion time for the whole project.

(b) Construct a resource histogram for the project showing the number of workers required each day, assuming each activity is to start as early as possible.

(c) Given that the group consists of 18 workers, explain why it would be necessary to reschedule some of the activities to enable them to finish the project in the shortest possible completion time. Give two possible ways in which this could be done.

(AQA)

12 A building project has been divided into a number of activities as shown in the table below.

	Activity	Immediate predecessor	Duration (days)
A	Prepare site	–	1
B	Erect walls	A	4
C	Excavate and lay drains	A	7
D	Fit window frames	B	3
E	Erect roof	B	9
F	Install electrics	B	7
G	Install plumbing	C, D	6
H	Insulate roof	E	3
I	Plaster walls	F, G, H	2

(a) Construct an activity network for the project
(b) Find the earliest start time for each activity.
(c) Find the latest finish time for each activity.
(d) Identify the critical path and state the shortest completion time for the whole project
(e) Construct a cascade (Gantt) diagram for the project, assuming each activity is to start as early as possible.
(f) Each activity requires one worker. Draw a resource histogram showing the number of workers required each day.
(g) Given that there are only three workers available on any day, show how the activities could be allocated so that the project may be finished in the shortest completion time.

(AQA)

13 The shell of a flat has been constructed and the inside is to be completed. The work has been divided into 13 activities, as shown in the table below.

	Activity	Tradesman	Immediate predecessors	Duration (days)
A	Studding	Builder	–	2
B	Electrics (first fix)	Electrician	A	1
C	Plumbing (first fix)	Plumber	A	1.5
D	Plaster boarding	Plasterer	B, C	1
E	Plastering	Plasterer	D	1
F	Artexing	Decorator	E	1
G	Electrics (second fix)	Electrician	E	0.5
H	Joinery (first fix)	Joiner	E	1
I	Plumbing (second fix)	Plumber	E	1
J	Joinery (second fix)	Joiner	G, H, I	2
K	Painting	Decorator	F, J	1
L	Tiling	Tiler	I, J	0.5
M	Cleaning	Cleaner	K, L	0.5

(a) Construct an activity network for the project.

(b) Find the earliest start time for each activity.

(c) Find the latest finish time for each activity.

(d) Identify the critical activities and state the shortest completion time for the whole project.

(e) Construct a cascade (Gantt) diagram for the project, assuming that each activity is to start as early as possible.

(f) The foreman of the project can speed up the project by using either an extra electrician, or an extra plumber or an extra joiner.

An extra electrician would reduce the time for electrical work by 50%, an extra plumber would reduce the time for plumbing work by 25%, and an extra joiner would reduce the time for joinery work by 20%.

Explain which extra tradesman the foreman should use, and find the new minimum time to complete the project. (AQA)

4 Dynamic programming

This chapter looks at a method of analysing optimisation problems involving sequences of decisions. When you have completed it you should

- understand the idea of working backwards with sub-optimisation
- know what is meant by stage and state variables, actions and costs and be able to use these ideas
- know how to set up dynamic programming tabulations
- be able to use dynamic programming tabulations to solve minimising problems.

4.1 Negative edge weights

In D1 Chapter 4, you studied how to solve shortest path problems using Dijkstra's algorithm. In that chapter it was noted that one drawback of Dijkstra's algorithm is that it cannot be used if any weights are negative. (The cost of a route might be negative if, for example, a firm will make a profit by making a delivery along that route.)

For example, consider the directed network shown in Fig. 4.1. You can see by inspection that the shortest path from A to C is ABC, which has weight 1. However, Dijkstra's algorithm permanently labels vertex C *before* route ABC is considered.

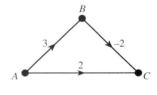

Fig. 4.1

An alternative method of solving shortest path problems, which can cope with negative edge weights is called **dynamic programming**. This is a term coined by Richard Bellman to describe techniques he and others developed in the 1950s to study optimisation problems involving sequences of decisions to be made by managers. Typically it is applied to production planning, where the task is to minimise the cost of a production process.

4.2 Terminology

Dynamic programming employs certain terminology which will be illustrated by the network of Fig. 4.2.

The vertices are called **states,** the directed edges are called **actions** and the weight on an edge is called the **cost**.

An action is what transforms a situation from one state to another.

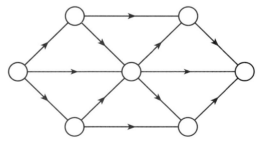

Fig. 4.2

If you look at Fig. 4.2, you can see that one state requires a maximum 4 actions before reaching the final state, two require 3 actions, one requires 2 actions, and so on. The final state requires 0 actions to get to the final state. The maximum number of actions to get from a state to the final state is called the **stage variable**. Fig. 4.3 shows the stage variables for each state.

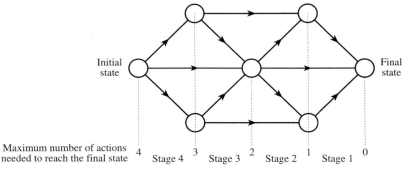

Fig. 4.3

The transition from a state to another whose stage variable is one less is called a **stage**. More specifically, the transition from states with stage variable k to states with stage variable $k-1$ is called Stage k. This is shown in Fig. 4.3.

Notice that the transition from a state with stage variable 4 to one with stage variable 2 is not a stage. Note also the importance of the word 'maximum' in the definition of stage variable. In Fig. 4.3, it is possible to get from the initial state to the final state with 2 actions, but the stage variable is 4 because this is the maximum number of actions required.

In addition to the stage variable, you can define a state variable for each state. For all the states with the same value of the stage variable the **state variable** is the number of the state as you move down the diagram from top to bottom.

Each state can be labelled with a pair of numbers, (stage variable, state variable). Fig 4.4 shows the labels for Fig. 4.3.

Thus the labelling of a state as $(3, 2)$ shows that

- the maximum possible number of actions to move to the final state is 3,
- this is the second such state moving down the diagram from top to bottom.

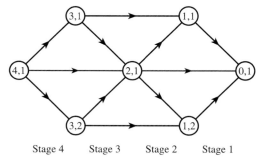

Fig. 4.4

Example 4.2.1
Label the states of Fig. 4.5a.

The solution is shown in Fig. 4.5b.

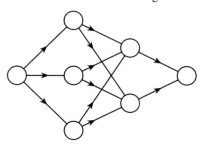

Fig. 4.5a

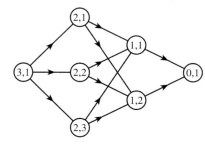

Fig. 4.5b

4.3 Minimising problems

Richard Bellman realised that a simple, general strategy can be used to solve optimisation problems involving a sequence of decisions. The principle of his method is that you should work backwards through the decisions, at each stage working out the best strategy from that point.

In this section you will see how to apply this idea to problems which require finding minima. The idea, when stated in a general way, is called Bellman's optimising principle.

> To solve an optimising problem involving a sequence of decisions you can work backwards. At each stage you should work out the best strategy from that point, called the **sub-optimal strategy**.

Suppose that to meet an increased demand for electricity in a town, it is intended to increase the capacity of parts of the electricity grid on one route from the generator to the town. The costs, in £10,000s, of bringing each part of the grid up to maximum capacity are shown in Fig. 4.6.

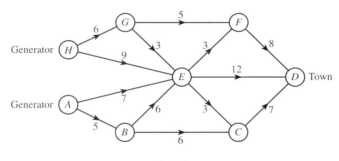

Fig. 4.6

To use dynamic programming to find the cheapest way of upgrading the grid, first label the states in the standard way. This gives

$$D(0,1), \quad F(1,1), \quad C(1,2), \quad E(2,1), \quad G(3,1), \quad B(3,2), \quad H(4,1), \quad A(4,2).$$

Similarly, number the actions from each state from top to bottom. Thus, focusing for the moment on state E, the actions EF, ED and EC are numbered 1, 2 and 3 respectively.

Alternatively, each action can be labelled by reference to the state to which it moves. For E, these labels would be $(1,1)$, $(0,1)$ and $(1,2)$.

Labelling the states (vertices) and numbering the actions (edges) in this way enables you to work systematically back through the network considering all possible courses of action. The working can be laid out in tabular form, as in Table 4.7. The work for two stages is shown, and a commentary follows the table.

Stage	State	Action	Value	Current minimum
1	$F(1,1)$	1	8	$\leftarrow 8$
	$C(1,2)$	1	7	$\leftarrow 7$
2	$E(2,1)$	1	$3+8=11$	
		2	12	
		3	$3+7=10$	$\leftarrow 10$

Table 4.7

The first two rows of Table 4.7 refer to Stage 1. During Stage 1, you know that it costs 8 to go from F to D, and 7 to go from C to D. As these are the only transitions from F and C to D, no choices need to be made; these are the minimum routes from F and C.

The next three rows refer to Stage 2; E is the only state with stage variable 2.

The first action from E is the transition to F. This route to D costs $3+8=11$.

The second action from E is the direct transition to D, which costs 12.

The third action from E is the transition to C. This route to D costs $3+7=10$.

The three routes from E to D cost 11, 12 and 10; the minimum is 10, so the value of 10 is used from now on for all routes through $E(2,1)$. Thus the sub-optimal strategy from E is the route with value 10.

The dynamic programming process is continued in Table 4.8.

Stage	State	Action	Value	Current minimum
1	$F(1,1)$	1	8	$\leftarrow 8$
	$C(1,2)$	1	7	$\leftarrow 7$
2	$E(2,1)$	1	$3+8=11$	
		2	12	
		3	$3+7=10$	$\leftarrow 10$
3	$G(3,1)$	1	$5+8=13$	$\leftarrow 13$
		2	$3+10=13$	$\leftarrow 13$
	$B(3,2)$	1	$6+10=16$	
		2	$6+7=13$	$\leftarrow 13$
4	$H(4,1)$	1	$6+13=19$	$\leftarrow 19$
		2	$9+10=19$	$\leftarrow 19$
	$A(4,2)$	1	$7+10=17$	$\leftarrow 17$
		2	$5+13=18$	

Table 4.8

The first row of Stage 3 refers to the route from G via F. This gives the cost of the action GF, which is 5, plus the minimum cost from F, which is 8, making a total of $5 + 8 = 13$.

The second row of Stage 3 refers to the route from G via E. This gives the cost of the action GE, which is 3, plus the minimum cost from E, which is 10, making a total of $3 + 10 = 13$.

Stage 4 is completed in a similar way. Because it is the final stage, the process ends having found minimum costs of 19 and 17, representing £19,000 from H and £17,000 from A.

It is therefore better to increase the capacity from Generator A. Retracing the best actions, namely action 1 from $A(4,2)$ to $E(2,1)$, then action 3 from $E(2,1)$ to $C(1,2)$ and finally action 1 from $C(1,2)$ to $D(0,1)$, yields the route $AECD$.

Example 4.3.1
Set up and use a dynamic programming tabulation to solve the problem considered in Section 4.1 of finding the shortest path from A to C in the first diagram.

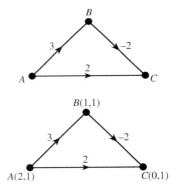

First label the states, shown in the second diagram.

Stage	State	Action	Value	Current minimum
1	$B(1,1)$	1	-2	$\leftarrow$ -2
2	$A(2,1)$	1	$3+(-2)=1$	$\leftarrow$ 1
		2	2	

The shortest path is ABC, of length 1.

You can solve a problem involving maximising in a very similar way.

Example 4.3.2*

The table shows the tasks in a project. Use dynamic programming to find the critical path.

Activity	Duration (days)	Immediate predecessors
A	10	–
B	11	A
C	6	–
D	10	A
E	7	B, C
F	9	D
G	5	E, F
H	4	G
I	7	E, F

Although it is possible to go straight from this information into a dynamic programming table without drawing an activity network, it is a good idea to start by drawing the network, and labelling the vertices using the dynamic programming conventions, with the slight modification that the activities (actions) are labelled with letters rather than numbers. Note, however, that the method of dealing with these actions is still systematic. For each state you consider the actions from top to bottom as if they were numbered.

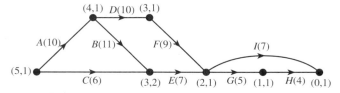

You are now in a position to set up the dynamic programming table.

Stage	State	Activity	Value	Current maximum
1	(1,1)	H	4	←4
2	(2,1)	I	7	
		G	5 + 4 = 9	←9
3	(3,1)	F	9 + 9 = 18	←18
	(3,2)	E	7 + 9 = 16	←16
4	(4,1)	D	10 + 18 = 28	←28
		B	11 + 16 = 27	
5	(5,1)	A	10 + 28 = 38	←38
		C	6 + 16 = 22	

The maximum duration, corresponding to the critical path, is 38 days. By retracing, the critical path is *ADFGH*.

Exercise 4

1. Set up stage and state variables for the vertices of this diagram.

2. The various possible routes for a person commuting by car are as shown in the diagram. The weights represent the number of traffic lights per leg of the journey.

   Use dynamic programming to find the route which minimises the number of traffic lights.

3. Given that the edge weights represent distances in miles, set up and use a dynamic programming tabulation to find the shortest route from A to E.

4. It is required to find the shortest route from A to D.
   (a) Explain why Dijkstra's algorithm is not an appropriate method for this problem.
   (b) Use dynamic programming to find the required route.

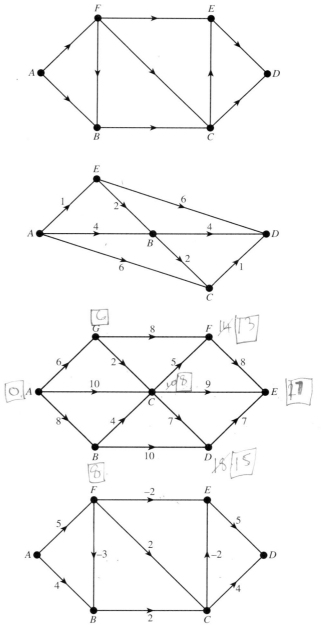

5* Set up and use a dynamic programming tabulation to find the longest paths from each of A, B and C to N.

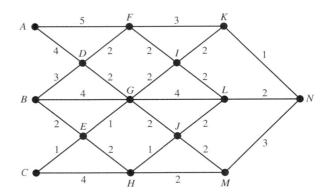

Miscellaneous exercise 4

1 A factory needs a particular raw material for part of its production. There are three suppliers, A, B and C of the raw material but their charges vary according to the month. The raw material from the three suppliers varies slightly and this necessitates a changeover cost to the factory. This cost depends on the supplier of the raw material during the previous month. During March the factory is set up for supplier A and at the end of June it is to return to this state.

The changeover costs, in pounds, are as follows.

		To		
		A	B	C
	A	0	20	24
From	B	16	0	16
	C	20	16	0

The costs, in pounds, of the raw material from the suppliers for the three months are as follows.

		Month		
		April	May	June
	A	60	70	50
Supplier	B	40	60	80
	C	20	40	60

(a) Complete a network to illustrate the possible ways that the factory may purchase the raw material during the three months.

(b) Find the total cost incurred if the factory purchases the raw material from supplier B during April, supplier C during May and supplier A during June.

(c) Use dynamic programming to find the shortest path through the network that corresponds to the minimum cost to the factory. Describe the purchasing plan that the factory should adopt and state the minimum cost.   (AQA, adapted)

2   A student is comparing the use of different algorithms for solving shortest path problems. For a graph with n vertices, the student obtains the following formulas for the numbers of operations needed:

   Dijkstra's algorithm $1.5n^2 - 2.5n + 1$,
   Dynamic programming $2n^3 - 9n^2 + 14n - 7$.

   (a) Hence comment on the relative merits of using Dijkstra's algorithm.

   (b) What advantage does dynamic programming have?

3   The network represents a set of routes, with the weights on the edges representing distances in metres. Use dynamic programming to find the shortest route from S to T. (You must demonstrate your use of the algorithm.)   (AQA, adapted)

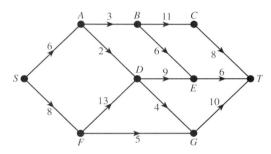

4   A company is planning three building projects, A, B and C, to be completed at the rate of one per year. The costs of each project depend on which of the other projects have already been completed, as given in the table.

   (a) Draw a network diagram to represent this information, using
         stage = projects still to be completed.

   (b) Find the building program for which the total cost is least.   (OCR, adapted)

Completed	Costs (£100,000s)		
	A	B	C
–	70	60	52
A	–	65	61
B	85	–	68
C	80	70	–
A, B	–	–	80
A, C	–	85	–
B, C	95	–	–

5   The network in the figure represents the direct routes between seven towns. The weights indicated on the edges record the time, in minutes, taken to travel the route.

   (a) Use dynamic programming starting at T, to determine the minimum total time to travel between towns S and T. In your solution, carefully identify the stages, states, actions and values.

   (b) Identify the route with the minimum time.

   (c) State one advantage and one disadvantage of using dynamic programming in this context rather than an algorithm such as Dijkstra's.   (OCR, adapted)

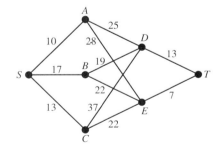

6 Each Saturday a speciality cake baker takes orders for the week ahead. One week the orders are as shown in the table.

Day cakes are required	Monday	Tuesday	Wednesday	Thursday	Friday
Number of cakes	1	2	5	3	2

The baker can make up to four cakes on any one day, and can store up to and including three cakes overnight. The cost of making each cake is £6, and the cost of storing a cake overnight is £0.50 (to be charged before storage). Cakes can be stored for more than one night. The daily overheads charge is £2.50 for each day that the baker is making cakes, but there is no overheads charge if no cakes are made. At the start and end of the week there must be no cakes in storage.

The baker wants to use dynamic programming to plan the production of the cakes so as to minimise the total costs.

The days will be the stages, the number of cakes in storage at the start of each day will be the states, and the number of cakes made each day will be the actions for this problem.

Set up a dynamic programming tabulation, working backwards from the end of the week, with columns for the stages, states, actions, number of cakes to be stored at the end of the day, daily costs and total costs. Use your tabulation to find the optimal production strategy.

(OCR)

7 The figure shows a network with costs on the edges.

Construct a dynamic programming tabulation showing stages, states, actions and a column for costs. Complete the table and find the route with the minimum total cost. (OCR, adapted)

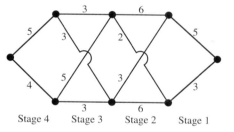

8* The Rolling Pebbles are choosing which venues to include on a short tour.

On Monday they will play at one of A, B and C; on Tuesday they will play at one of D, E and F; on Wednesday they will play at one of G, H and I; they return home on Thursday.

The figure shows the venues, labelled using (stage, state). The weights on the edges show the expected profit (in £10,000) from playing at each venue.

Use dynamic programming, working backwards from Thursday, to find which venues should be played on which day to maximise the total profit.

(OCR, adapted)

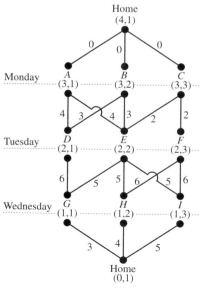

9 A theatre company intends to perform three plays, A, B and C, during a particular season. The order in which they perform the plays is a matter of choice. However, because of the need to construct and dismantle scenery, the running cost for each play depends on which, if any, plays have preceded it. The costs in thousands of pounds are given in the table.

Play number	Previous play(s)	Cost (£000s)		
		A	B	C
1	–	11	9	7
2	A	–	10	9
	B	14	–	10
	C	12	12	–
3	A & B	–	–	12
	A & C	–	13	–
	B & C	14	–	–

Use dynamic programming to determine an order of plays that minimises the total cost for the season.

(AQA)

5 The Simplex algorithm

This chapter is about solving linear programming problems when graphical methods are not easily available. When you have completed it you should

- be able to solve linear programming problems by the Simplex algorithm.

5.1 The Simplex method

The graphical method you have used in D1 Chapter 8 is an excellent method when it works. However, it is necessary to be able to solve problems without drawing a diagram. The Simplex method has been designed to do just this. Such a method is of course essential for solving problems on a computer. It is also essential if you have a practical problem with lots of variables which cannot satisfactorily be represented by a two-dimensional drawing.

You may wish to refer to D1 Chapter 8 for a reminder about the terminology.

Consider the opening question of D1 Chapter 8, which was

$$\text{maximise} \quad P = x + y,$$

$$\text{subject to} \quad 2x + y \leq 16,$$
$$2x + 3y \leq 24,$$
$$x \geq 0, \, y \geq 0.$$

To use the Simplex method, you first introduce **slack variables**, s and t say, to convert the two non-trivial inequalities into equalities.

Let $\quad s = 16 - 2x - y \qquad\qquad$ Equation 1

and $\quad t = 24 - 2x - 3y \qquad\qquad$ Equation 2

Then the inequalities reduce to $x \geq 0, \, y \geq 0, \, s \geq 0, \, t \geq 0$.

The vertices of the feasible region (see Fig. 5.1) then correspond to places where two of the four variables are zero. Check, for example, that A is the point where $y = 0$ and $s = 0$.

The points O, A, B and C are important because, as you have seen, the objective function will be maximised at one of them.

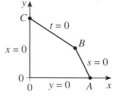

Fig. 5.1

Equations 1 and 2 can be used to write the objective function P in terms of any pair of variables. For example, writing x as $P - y$ and substituting in Equation 1 you get

$$s = 16 - 2(P - y) - y, \quad \text{which reduces to} \quad s = 16 - 2P + 2y - y = 16 - 2P + y.$$

Making P the subject of the equation $s = 16 - 2P + y$ gives

$$P = 8 + \tfrac{1}{2}y - \tfrac{1}{2}s.$$

You should check that you can obtain the alternatives

$$P = x + y, \quad P = 8 + \tfrac{1}{2}y - \tfrac{1}{2}s, \quad P = 10 - \tfrac{1}{4}s - \tfrac{1}{4}t, \quad P = 8 + \tfrac{1}{3}x - \tfrac{1}{3}t.$$

The third expression, $P = 10 - \tfrac{1}{4}s - \tfrac{1}{4}t$, is special because the coefficients of the two variables, s and t, are both negative. Since $s \geq 0$ and $t \geq 0$ this means that P *cannot* be greater than 10. The maximum is therefore 10, occurring when $s = t = 0$, which is at the point where $x = 6$ and $y = 4$.

This method seems much clumsier than the graphical method! However, it can be speeded up by adopting a clever notation for the linear equations which need to be manipulated to find the expressions for P. Moreover, it can be programmed for many variables.

5.2 The tableau format

Consider solving the two linear equations

$$\begin{aligned} x + 2y &= 7, \quad &\text{Equation 1} \\ 3x - 4y &= -9. \quad &\text{Equation 2} \end{aligned}$$

It is legitimate to double an equation:

$$2x + 4y = 14. \quad \text{Equation 3} = 2 \times \text{Equation 1}.$$

At this stage it is useful to have a shorter notation for Equation 1 and Equation 2. A number in a circle, such as ①, will mean Equation 1. Thus the previous line, which was explained by Equation 3 = 2 × Equation 1, would now be explained by ③ = 2 × ①.

Returning to sets of equations, it is also legitimate to add (or subtract) them.

Thus adding ② and ③ gives

$$5x = 5. \quad ④ = ② + ③$$

Dividing this equation by 5, you get

$$x = 1. \quad ⑤ = \tfrac{1}{5} \times ④$$

Continuing, you should obtain $y = 3$.

You can write these operations neatly in a tableau, shown in Table 5.2.

x	y		Equation	
1	2	7	①	
3	−4	−9	②	
2	4	14	③ = 2 × ①	
5	0	5	④ = ② + ③	
1	0	1	⑤ = $\tfrac{1}{5}$ × ④	This equation is just $x = 1$.
0	2	6	⑥ = ① − ⑤	
0	1	3	⑦ = $\tfrac{1}{2}$ × ⑥	This equation is $y = 3$.

Table 5.2

The tableau format becomes even more useful as the number of variables increases. Consider, for example, the manipulation of equations which occurred in Section 5.1.

Example 5.2.1
Given that $P = x + y$, $2x + y + s = 16$ and $2x + 3y + t = 24$, use a tableau to express P in terms of s and t.

P	x	y	s	t		Equation	
1	−1	−1	0	0	0	①	
0	2	1	1	0	16	②	
0	2	3	0	1	24	③	
1	0	$-\frac{1}{2}$	$\frac{1}{2}$	0	8	④ = ① + $\frac{1}{2}$ × ②	Equation ② is used
0	0	2	−1	1	8	⑤ = ③ − ②	to eliminate x.
1	0	0	$\frac{1}{4}$	$\frac{1}{4}$	10	⑥ = ④ + $\frac{1}{4}$ × ⑤	Then y is eliminated.

So $P + \frac{1}{4}s + \frac{1}{4}t = 10$, giving $P = 10 - \frac{1}{4}s - \frac{1}{4}t$, as before.

> In the tableau format, rows can be
> - multiplied by constants or divided by (non-zero) constants,
> - added or subtracted.

5.3 The Simplex algorithm

The ideas of Sections 5.1 and 5.2 can now be combined to give you a general (and easily programmed) method of solving a linear programming **maximising** problem.

The objective function must be expressed using an equation in which the right side is a number. For example, if $P = 7x + 11y$, then this is written as $P - 7x - 11y = 0$. Each non-trivial constraint must also be expressed as an equation by using slack variables. For example, $2x + 4y \leq 60$ would be written as $2x + 4y + s = 60$, where $s \geq 0$, and $2x + 4y \geq 60$ would be written as $2x + 4y - s = 60$, where $s \geq 0$.

The shaded box on the next page describes the Simplex algorithm, and the example after it illustrates how to do the algorithm in practice.

A commentary on the whole process follows after the tableau. You are advised to read the algorithm and study the tableau and commentary together.

For $2x + 4y + s = 60$, s is the 'slack' in the inequality $2x + 4y \leq 60$. For example, if $x = 6$ and $y = 10$, then $2x + 4y = 52$, and the slack is $s = 8$.

The Simplex algorithm

Step 1 Formulate the maximising problem (using slack variables as necessary) as a tableau.

Step 2 Ensure that all elements in the last column (except possibly the top one) are non-negative.

Step 3 Select any column (except the last one) whose top element is negative.

Step 4 Call the numbers in the selected column a_0, a_1, a_2, ... and the numbers in the last column l_0, l_1, l_2, ... In the selected column, choose the positive element a_i for which $\dfrac{l_i}{a_i}$ is least. This a_i is called the **pivot**.

Step 5 Divide the ith row by a_i.

Step 6 Combine appropriate multiples of the ith row with all the other rows in order to reduce to zero all other elements in the column of the pivot.

Step 7 If all top elements (except possibly the last one) are non-negative then the maximum has been reached. Otherwise return to Step 3.

Step 8 The last column contains the values of the objective function and the non-zero variables.

Example 5.3.1

Maximise $7x + 11y$ subject to the constraints $2x + 4y \leq 60$, $3x + 3y \leq 60$ and $x, y \geq 0$.

Start by introducing the slack variables s and t, by writing $2x + 4y + s = 60$ and $3x + 3y + t = 60$, so that $s \geq 0$ and $t \geq 0$. Remember also that $P = 7x + 11y$ must be written in the form $P - 7x - 11y = 0$.

In the tableau, the pivots are shown in bold type. You are advised to *ring* the pivots.

P	x	y	s	t	l	Equation	
1	−7	−11	0	0	0	①	
0	2	4	1	0	60	②	
0	**3**	3	0	1	60	③	3 is the pivot in the x column.
1	0	−4	0	$\frac{7}{3}$	140	④ = ① + $\frac{7}{3}$ × ③	
0	0	**2**	1	$-\frac{2}{3}$	20	⑤ = ② − $\frac{2}{3}$ × ③	2 is the pivot in the y column.
0	1	1	0	$\frac{1}{3}$	20	⑥ = $\frac{1}{3}$ × ③	
1	0	0	2	1	180	⑦ = ④ + 2 × ⑤	
0	0	1	$\frac{1}{2}$	$-\frac{1}{3}$	10	⑧ = $\frac{1}{2}$ × ⑤	
0	1	0	$-\frac{1}{2}$	$\frac{2}{3}$	10	⑨ = ⑥ − $\frac{1}{2}$ × ⑤	

The maximum value is 180, when $x = 10$ and $y = 10$.

Here is a commentary on the solution to Example 5.3.1. For Steps 1 and 2, the original equations are written as the first three rows of the tableau.

Step 3   The x-column is chosen as one which has a negative top element.

Step 4   In the x-column, $a_0 = -7$ so it can't be the pivot. Looking at the others, $a_1 = 2$, $l_1 = 60$ so $\frac{l_1}{a_1} = 30$; $a_2 = 3$, $l_2 = 60$ so $\frac{l_2}{a_2} = 20$. As 20 is the least, $a_2 = 3$ is the pivot.

Step 5   Equation 3 is then divided by the pivot, and labelled Equation 6. (It is put in this position so that the essential order of Equations 1, 2 and 3 is maintained.)

Step 6   Then Equation 4 and Equation 5 are derived from Equation 1 and Equation 2 with their x-coefficients made 0 by subtracting appropriate multiples of Equation 3.

When you have derived Equations 4, 5 and 6, you have completed Step 6 and carried out one iteration of the Simplex algorithm. It is conventional to rule a line under the equations at this stage.

The top equation is now $P - 4y + \frac{7}{3}t = 140$, where $y \geq 0$ and $t \geq 0$, so you can make P larger by increasing y.

Step 7   At this stage not all the elements in the top row (Row 4) are non-negative: there is an element -4. So you have to return to Step 3 and choose another column.

Step 3   The y-column is chosen, as the number at the top, -4, is negative.

Step 4   This time 2 is the pivot, because $\frac{20}{2} = 10$ is less than $\frac{20}{1} = 20$, and -4 is not a candidate for the pivot because it is negative.

Step 5   Equation 5 is then divided by the pivot, and becomes Equation 8, in the same position as Equation 5.

Step 6   Then Equation 7 and Equation 9 are derived from Equation 4 and Equation 6, with their y-coefficients made 0 by subtracting appropriate multiples of Equation 5.

Step 7   All the top elements (that is, those in Equation 7) are now non-negative, so you move to Step 8.

Step 8   At this stage, Equation 7, which is $P + 2s + t = 180$, tells you that the maximum value of the objective function is 180. This occurs when s and t both take the value 0. Equation 8, which is $y + \frac{1}{2}s - \frac{1}{3}t = 10$, tells you that this happens when $y = 10$. Equation 9, which is $x - \frac{1}{2}s + \frac{2}{3}t = 10$, tells you that $x = 10$.

The Simplex algorithm can be used to minimise a function with a simple trick: to minimise a linear function, you maximise its negative. Thus to minimise $4x - 5y - 3z$, you maximise $-4x + 5y + 3z$. The minimum for the original problem is then the negative of the maximum.

Example 5.3.2
Minimise $4x - 5y - 3z$ subject to $x - y + z \geq -2$, $x + y + 2z \leq 3$ and $x, y, z \geq 0$.

Start by introducing the slack variables s and t by writing $x - y + z - s = -2$ and $x + y + 2z + t = 3$, where $s \geq 0$ and $t \geq 0$. Recall that in the tableau the elements in the last column (except possibly the top one) must not be negative. Therefore the first equation is written in the form $-x + y - z + s = 2$.

Next, let $P = -4x + 5y + 3z$, since you are going to minimise $4x - 5y - 3z$ by maximising its negative. The equation $P = -4x + 5y + 3z$ must be written in the form $P + 4x - 5y - 3z = 0$.

Here is the tableau.

P	x	y	z	s	t		Equation
1	4	−5	−3	0	0	0	①
0	−1	**1**	−1	1	0	2	②
0	1	1	2	0	1	3	③
1	−1	0	−8	5	0	10	④ = ① + 5× ②
0	−1	1	−1	1	0	2	⑤ = ②
0	2	0	3	−1	1	1	⑥ = ③ − ②
1	$\frac{13}{3}$	0	0	$\frac{7}{3}$	$\frac{8}{3}$	$\frac{38}{3}$	⑦ = ④ + $\frac{8}{3}$ × ⑥
0	$-\frac{1}{3}$	1	0	$\frac{2}{3}$	$\frac{1}{3}$	$\frac{7}{3}$	⑧ = ⑤ + $\frac{1}{3}$ × ⑥
0	$\frac{2}{3}$	0	1	$-\frac{1}{3}$	$\frac{1}{3}$	$\frac{1}{3}$	⑨ = $\frac{1}{3}$ × ⑥

The maximum value of $-4x + 5y + 3z$ is $\frac{38}{3}$, or $12\frac{2}{3}$, so the minimum value of $-4x + 5y + 3z$ is $-12\frac{2}{3}$. This occurs when $x = 0$, $y = \frac{7}{3}$ and $z = \frac{1}{3}$.

The reasons for the values of x, y and z come from examining Equations 7, 8 and 9 in conjunction with the inequalities $x, y, z, s, t \geq 0$. Equation 7 is $P + \frac{13}{3}x + \frac{7}{3}s + \frac{8}{3}t = \frac{38}{3}$. This shows that the maximum value of P occurs when $x = s = t = 0$. Equation 8 is $-\frac{1}{3}x + y + \frac{2}{3}s + \frac{1}{3}t = \frac{7}{3}$, and since $x = s = t = 0$ this gives $y = \frac{7}{3}$. Similarly $z = \frac{1}{3}$.

Note that using fractions in the Simplex tableau avoids the danger of rounding errors. Although when a computer is programmed for the Simplex algorithm it works in decimals, to a large number of places to retain accuracy, you should use fractions.

In theory, it is possible for the Simplex algorithm to get caught in an infinite loop. However, in practice, such a difficulty is extremely rare.

But there is a problem which can arise at the start of the method. As you have seen, the idea of the Simplex method is to start at the point where all the control variables are zero and then move to 'better' points. However, suppose that one constraint was $x - y \geq 1$. Then the point where $x = 0$ and $y = 0$ is not even in the feasible region, and the process cannot start. Methods have been developed to deal with this kind of difficulty, but they are beyond the scope of this book.

Exercise 5A

1. (a) Maximise $2x + 3y$ subject to the constraints $x + 2y \leq 6$, $x + y \leq 5$ and $x, y \geq 0$ by using the Simplex method. Give the whole tableau in your answer.

   (b) Repeat part (a) with the objective function $2x + y$.

2. Use the Simplex algorithm to maximise $x + 4y$ subject to $x + 3y \leq 15$, $2x + y \leq 12$ and $x, y \geq 0$. Show the whole tableau in your answer.

3. Maximise $5x + 3y - 8z$ subject to the constraints $x - z \leq 1$, $x + y \leq 2$, $3x + 2y - 4z \leq 6$ and $x, y, z \geq 0$. Show the whole Simplex tableau.

4. Three processes, I, II and III, are involved in the manufacture of three products, A, B and C. For each product, the manufacturing time (hours) and profit per item (£) are as shown.

Product	I	II	III	Profit
A	1	2	3	120
B	5	1	2	70
C	4	4	1	100

   The total available manufacturing times on processes I, II and III are 90, 35 and 60 hours respectively.

   (a) What mix of products yields the greatest profit?

   (b) What assumptions have you made in answering part (a)?

5. The marketing director of an insurance company has a budget of £100,000 for half-page advertisements in three specialist magazines. Details of the circulations and advertising costs of these magazines are as shown.

Magazine	Circulation	Cost (£)
Smart Savings	80 000	1000
Capital Investor	120 000	2000
Money Monthly	200 000	2500

   The director is instructed that the average circulation for each advertisement should be at least 100 000 and that no more than £40,000 should be spent on any one magazine. What is the maximum number of advertisements she can place and how can she achieve this?

6. Minimise $x + y - 2z$ subject to $2x + y \geq z$, $2x + 5 \geq 3z$, $3y + 4z \leq 12$ and $x, y, z \geq 0$.

5.4 Network problems

You have seen how linear programming is concerned with making something optimal (for example, least costs or greatest profit) subject to various constraints. In D1 you saw a range of network problems which were also about constrained optimisation. Each of these problems can actually be converted into a linear programming problem.

Here is a very simple example of the shortest path problem.

Example 5.4.1   Shortest paths
Find the shortest path from vertex A to vertex C in Fig. 5.3.

To convert this into a linear programming question, define a, b and c by

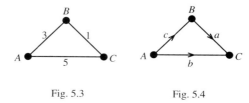

Fig. 5.3   Fig. 5.4

$$a = \begin{cases} 1 & \text{if } BC \text{ is on the shortest path} \\ 0 & \text{otherwise.} \end{cases}$$

The quantities b and c are defined similarly: see Fig. 5.4.

The objective function is then the total length of the chosen edges, that is

$a + 5b + 3c$.

Here is the linear programming formulation of the problem.

Minimise $a + 5b + 3c$.

Constraints $b + c = 1$,   There is an edge out of A.
$a + b = 1$,   There is an edge into C.
$a = c$,   There is an edge out of B if there is an edge into B.
$a, b, c \geqslant 0$.

You might well think that each of the final three conditions should take the form $a = 0$ or $a = 1$ rather than $a \geqslant 0$. However, it can be proved that the given conditions are sufficient and will actually force each variable to be 0 or 1. Before reading on, solve the linear programming problem and obtain the obvious solution $a = c = 1$, $b = 0$.

Example 5.4.2   Minimum connector
Find the minimum connector for the network of Fig. 5.5.

Refer to Fig. 5.6, and define

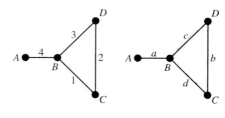

$$a = \begin{cases} 1 & \text{if } AB \text{ is used} \\ 0 & \text{otherwise.} \end{cases}$$

The quantities b, c and d are defined similarly.

Fig. 5.5   Fig. 5.6

Here is the problem in a linear programming format.

Minimise $\quad 4a + 2b + 3c + d.$

Constraints
$\quad a \geq 1,\quad$ (A is connected)
$\quad a + c + d \geq 1,\quad$ (B is connected)
$\quad b + d \geq 1,\quad$ (C is connected)
$\quad b + c \geq 1,\quad$ (D is connected)
$\quad c + d \geq 1,\quad$ (A and B are connected)
and so on, and
$a, b, c, d \geq 0.$

The constraints omitted above do not add anything new. For example, $a + b + c \geq 1$, which arises because B and C are connected, is already covered by $a \geq 1$ and $b + c \geq 1$.

Again, before reading on, you should solve the linear programming problem. You will obtain the obvious solution $a = b = d = 1, c = 0$.

As you can see, linear programming provides a general problem-solving technique which can be applied to the various network problems studied in D1. Nevertheless, the specialised algorithms such as Dijkstra's, Prim's and Kruskal's are usually far easier to apply when dealing with a specific network problem.

Exercise 5B

1 The tableau format for a problem is as shown.

P	x	y	z	r	s	t	
1	−6	−5	−3	0	0	0	0
0	7	7	4	1	0	0	23
0	5	6	2	0	1	0	16
0	4	8	−2	0	0	1	13

(a) Identify the objective function, constraints, control variables and slack variables.

(b) Find the maximum possible value of the objective function and the corresponding values of the control variables.

2 Consider this linear programming problem.

Maximise $3x + 2y$ subject to the constraints $x + y \leq 40$, $2x + y \leq 50$, $-x + 4y \geq 20$ and $x, y \geq 0$.

(a) Use a graphical method to solve this problem.

(b) Minimise $3x + 2y$ subject to the same constraints.

(c) What difficulty would arise if you attempted to use the Simplex algorithm to tackle either part (a) or part (b)?

(d) Replace the second constraint by $2x + y \geq c$. Plot a graph of the maximum value of $3x + 2y$ against c.

3   A company is producing three models of personal computer for the Christmas market, the Supremo 64, the Supremo 128 and the Supremo 256. Each computer requires a standard microprocessor of which 3000 can be purchased and marketing information has led to the decision that at least five-sixths of the computers should be Supremo 64s. The company can raise £2,000,000 for purchasing components and has 5000 assembly hours available.

Details of assembly times, costs of components and selling prices are as shown in the table.

PC	Assembly time (hrs)	Cost (£)	Selling price (£)
64	1.5	600	1000
128	2	700	1200
256	3	800	1600

Advise the company as to the production strategy which will optimise profits. What assumptions have you made in formulating your advice?

4   (a) Display the following linear programming problem in a Simplex tableau.
    Maximise $P = 3x + 3y + 2z$,
    subject to $3x + 7y + 2z \leq 15$,
                $2x + 4y + z \leq 8$,
                $x \geq 0, y \geq 0, z \geq 0$.

   (b) Perform one iteration of the Simplex algorithm, choosing to pivot on an element from the x-column.

   (c) State the values of x, y, z and P resulting from the first iteration. How do you know whether or not you have found the optimal solution?

   (d) Complete the solution to this problem.

5* Consider the various linear programming problems which you have solved using the Simplex method. For any such problem let n be the number of variables (both control and slack), let m be the number of non-trivial constraints and let v be the number of non-zero variables (both control and slack) in the optimum solution.

Construct a table of values of n, m and v. What do you notice?

Miscellaneous exercise 5

1   Consider the linear programming problem:
    maximise $P = 4x + 5y + 3z$,
    subject to $8x + 5y + 2z \leq 4$,
                $x + 2y + 3z \leq 1$.

   (a) Set up a Simplex tableau for this problem.

   (b) Choose as first pivot an element in the x-column. Identify the values of the variables and the objective function after the first iteration.

   (c) Complete the Simplex method, explaining how you know that your solution is optimal.

2 A problem is formulated as maximise $P = 4000x + 6000y$ subject to $x + y \leq 5000$, $-x + 4y \leq 0$, $x + 3y \leq 6000$, $x \geq 0$ and $y \geq 0$.

  (a) By introducing slack variables, set up the problem as an initial Simplex tableau.

  (b) Perform one iteration of the tableau, beginning by choosing to pivot on the x-column. Interpret the result of this iteration.

  (c) Explain how you know, from the tableau, that the optimal solution has not yet been reached. (OCR, adapted)

3 A company manufactures three kinds of robot, the Brainy, the Superbrainy and the Superbrainy X. After production, each robot is tested for its coordination and its logic. The company wants to maximise its profit from the sale of the robots. You may assume that every robot manufactured is sold. The table below shows the times required by these tests, the time available each week, and the profit per robot.

	Coordination test (hours)	Logic test (hours)	Profit per robot (£)
Brainy	6	3	2400
Superbrainy	3	4	3000
Superbrainy X	1	5	3200
Time available (hours)	60	60	

  (a) Set up this problem as a linear programming problem.

  (b) Convert this problem into an initial Simplex tableau.

  (c) Perform one iteration of the tableau, explaining your method carefully. Interpret your solution in terms of the original problem. (OCR, adapted)

4 Phil decides to use an interlocking shelving system. The shelving system can be made up from any combination of narrow shelves, medium shelves and wide shelves.

Suppose that Phil buys x metres of narrow shelves, y metres of medium shelves and z metres of wide shelves.

The problem of maximising the storage capacity, when using the shelving system, gives the LP formulation.

  Maximise $P = 3x + 4y + 5z$,
  subject to $2x + 8y + 5z \leq 3$,
  $9x + 3y + 6z \leq 2$,
  and $x \geq 0, y \geq 0, z \geq 0$.

  (a) Set up an initial Simplex tableau for this problem. Perform two iterations, choosing to pivot first on an element chosen from the z column.

  (b) Interpret the result of each of the two iterations carried out in part (a).

  (c) Explain how you know whether or not the optimal solution has been achieved. (OCR)

5 Consider the linear programming problem below.

Maximise $P = 2x + y$,
subject to $x + y \leq 7$,
$x + 2y \leq 10$,
$2x + 3y \leq 16$,
$x \geq 0$ and $y \geq 0$.

(a) By introducing slack variables, represent the problem as an initial Simplex tableau.

(b) Perform one iteration of the Simplex algorithm, choosing to pivot first on an element chosen from the x-column.

(c) State the values of x, y and P resulting from the iteration in part (b).

(d) Explain how you know whether or not the optimal solution has been achieved. (OCR)

6* Consider the following linear programming problem.

Maximise $P = 3x + y$,
subject to $x + 3y \geq 10$,
$y + 5 \leq 2x$,
$3x + 2y \leq 20$,
and $x \geq 0,\ y \geq 0$.

(a) Graph the constraints, using $0 \leq x \leq 10$ and $0 \leq y \leq 10$. Identify the feasible region by shading those regions where the constraints are not satisfied.

(b) Draw and clearly label the line $P = 25$ on your graph.

(c) Use your graph to find the values of x and y that solve the linear programming problem, giving your answers correct to 1 decimal place.

The linear programming problem can be written in the form below (you are not expected to demonstrate this).

Maximise $R = 8X + 6Y + 900$,
subject to $-7X + Y + S = 0$,
$7X + 9Y + T = 350$,
and $X \geq 0,\ Y \geq 0,\ S \geq 0,\ T \geq 0$,

where $R = 70P$, $X = 7(3x - y) - 60$, $Y = 7(x + 3y) - 70$, and S and T are slack variables.

(d) Use the Simplex algorithm to calculate the optimum value of R, and the values of X and Y for which it is attained, choosing to pivot first on an element in the X column. Show sufficient working to make your method clear.

(e) Use your answers from part (d) to calculate the optimum value of P, and the values of x and y for which it is attained for the linear programming problem given at the start of the question. (OCR)

6 Game theory

This chapter looks at the analysis of situations where two competitors are both making decisions which affect each other. When you have completed it you should

- know what a two-person zero-sum game is
- be able to determine a play-safe strategy
- know what is meant by the value of a game with a stable solution
- be able to simplify games by using dominance arguments
- be able to find optimal mixed strategies for a game with no stable solution
- know how to convert a two-person zero-sum game into a linear programming problem.

6.1 Zero-sum games

In D1 and the previous chapters of this book, you have considered a variety of techniques of discrete mathematics. As you have seen, most of these methods were developed to address business or military applications.

However, in business or military applications not only are you making decisions but so are your competitors or potential enemies. So far this important aspect of discrete mathematics has been ignored and it is the purpose of this final chapter to remedy this omission.

The general idea of studying mathematically the consequences of decisions being made simultaneously by two or more competitors is called **game theory**. This theory has become a vital component of studies of competitive behaviour ranging from general interpersonal relationships to peace negotiations. In this chapter you will study the particular case of what is called a **two-person zero-sum game**.

The 'two-person' aspect has an obvious meaning: you will be considering cases where there are just two competitors. The 'zero-sum' aspect means that you will be studying cases where the gain for one party is exactly matched by the loss for the other. In zero-sum games you cannot achieve the ideal of a 'win-win' situation.

The idea of two-person zero-sum games can be illustrated by considering the game of 'stone-scissors-paper'. In this game, the two players simultaneously show either a fist representing stone, or two fingers representing scissors or an open palm representing paper.

The winner is determined by the following rules.
- Stone blunts and therefore beats scissors.
- Scissors cut and therefore beat paper.
- Paper wraps around and therefore beats stone.

If both players show the same object then the game is drawn.

The possible outcomes of one game can be represented by what are called **pay-off matrices**, shown in Fig. 6.1.

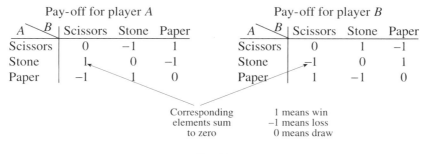

Fig. 6.1

The 'zero-sum' nature of this game is demonstrated by the fact that each pair of corresponding elements has a sum of zero. Since one array of pay-offs is simply the negative of the other it is conventional to write a single pay-off matrix for a game, the one corresponding to the player (A in Fig. 6.1) whose choices are written on the left side of the matrix.

> A **two-person zero-sum game** is one between two players such that whatever one player gains the other loses.
>
> A **pay-off matrix** for such a game is a rectangular array of numbers representing the outcomes of each possible pair of decisions. The **outcomes** are the pay-offs to the player whose choices are on the left side of the matrix.

6.2 Play-safe strategies

Consider a game with pay-offs as shown in Fig. 6.2.

If player A chooses option 1 then A may lose as much as £1M. Whereas choosing option 2 guarantees that A will at least break even. In practice, many players would ignore the riskier option 1 and 'play safe' with option 2.

		Player B	
		1	2
Player A	1	£3M	−£1M
	2	£0	£1M

Fig. 6.2

> A **play-safe** strategy is to choose the option whose worst outcome is as good as possible.

When determining play-safe strategies, the working can be laid out as in the next two examples.

Example 6.2.1
Two players, A and B, play a game with the pay-off matrix shown.
(a) What is the outcome if both players play safe?
(b) Can either player improve their strategy if they know the other will play safe?

A \ B	1	2
1	3	−4
2	−3	1
3	2	−2

(a) Put an extra column to the right to show the worst outcome for A in each row. Thus, if A plays option 1, A may gain 3 or gain ± 4; the worst outcome is gaining ± 4, that is losing 4. Similarly the worst outcomes for options 2 and 3 are -3 and -2. Thus the best of the worst outcomes is -2, indicated by the arrow. This is A's play-safe strategy.

Similarly an extra row at the bottom shows B's worst outcomes, which are losing 3 and 1. The best of these worst outcomes is 1, indicated by the arrow.

A \ B	1	2	Worst outcome for A
1	3	−4	−4
2	−3	1	−3
3	2	−2	−2 ←
Worst outcome for B	3	1	

Thus, playing safe means A chooses option 3 and B chooses option 2. The outcome is that A loses 2 and B gains 2.

(b) If B plays safe, A can improve the pay-off to $+1$ by choosing option 2.
If A plays safe, B can do no better than choose option 2.

Example 6.2.2
For a game with the pay-off matrix in the figure, find the outcome if both players adopt play-safe strategies. Show that neither player can improve on this outcome by adopting a different strategy.

$$\begin{pmatrix} -1 & 0 & 1 \\ -2 & 1 & 3 \\ -3 & 0 & -4 \end{pmatrix}$$

The pay-off matrix is now given without explanatory notation about who is winning and who is losing. This will now be the convention in exercises and examples.

The table shows the play-safe strategies. So A chooses option 1 for a gain of at least -1 and B chooses option 1 for a gain of at least 1. A change of option for A would lose 2 or 3 and a change for B would break even or lose 1. It is in both players' interests to adopt the play-safe strategy.

A \ B	1	2	3	Worst outcome for A
1	−1	0	1	−1 ←
2	−2	1	3	−2
3	−3	0	−4	−4
Worst outcome for B	−1	1	3	

From here on the extra row and column in the table will be labelled 'Row min(imum)' and 'Col(umn) max(imum)' instead of 'Worst outcome for A' and 'Worst outcome for B' respectively.

6.3 Stable solutions

The two examples of Section 6.2 illustrate that adopting a play-safe strategy may or may not be advantageous for a player.

If, as in Example 6.2.2, there is no incentive for either player to change from a play-safe strategy, then the game is said to have a **stable** solution.

Consider again the stable solution of Example 6.2.2, shown in Fig. 6.3.

You will have noticed that under this solution:

A can guarantee to win at least $\max(\text{row min}) = -1$;

B can guarantee *A* will win at most $\min(\text{col max}) = -1$.

The crucial result is that, for this matrix, the maximum of the row minima is equal to the minimum of the column maxima.

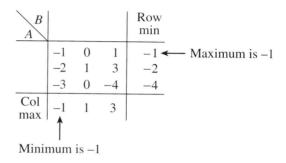

Fig. 6.3

A two-person zero-sum game has a **stable solution** if and only if

$$\max(\text{row minima}) = \min(\text{column maxima}).$$

If a game has a stable solution, then both players should adopt a play-safe strategy and the pay-off from these strategies is called the **value** of the game. The position in the pay-off matrix where this value occurs is called the **saddle point**.

Example 6.3.1

Determine whether the following games have stable solutions and, if so, state the value of the game.

(a) $\begin{pmatrix} 0 & -1 & 1 \\ 1 & 0 & -1 \\ -1 & 1 & 0 \end{pmatrix}$ (b) $\begin{pmatrix} -2 & 1 & 2 \\ 5 & 2 & 6 \\ 1 & -3 & 3 \end{pmatrix}$

(a) Each row minimum $= -1$, so $\max(\text{row min}) = -1$.

Each column maximum $= +1$, so $\min(\text{col max}) = 1$.

As $\max(\text{row min}) \neq \min(\text{col max})$, there is no stable solution.

(b) The row minima are -2, 2 and -3, so $\max(\text{row min}) = 2$.

The column maxima are 5, 2 and 6, so $\min(\text{col max}) = 2$.

As $\max(\text{row min}) = \min(\text{col max})$, the game has a stable solution. The value of the game is the common value of $\max(\text{row min})$ and $\min(\text{col max})$, 2.

One idea which you can sometimes use to simplify the process of finding a solution to a problem is that of **dominance**.

Consider, for example, a game with the pay-off matrix in Fig. 6.4.

A \ B	1	2	3	4
1	5	−4	0	5
2	4	3	1	3
3	6	2	−2	3

Fig. 6.4

B should never choose options 1 and 4 because each element of column 1 and of column 4 is greater than or equal to the corresponding element of column 2. Thus the minimum of columns 1, 2 and 4 is always the element in column 2; column 2 is said to 'dominate' column 1 and column 4. Therefore you only need to consider the matrix in Fig. 6.5.

A \ B	2	3
1	−4	0
2	3	1
3	2	−2

Fig. 6.5

Similarly, in Fig. 6.5, each element of row 3 is less than the corresponding element of row 2, so A should never choose option 3. This leaves just the 2×2 matrix in Fig. 6.6 to be considered.

A \ B	2	3
1	−4	0
2	3	1

Fig. 6.6

In fact, this has a stable solution when A chooses option 2 and B chooses option 3. The game therefore has value 1.

> A row of a pay-off matrix can be ignored if each element is less than or equal to the corresponding element of another row.
>
> A column of a pay-off matrix can be ignored if each element is greater than or equal to the corresponding element of another column.
>
> This process of ignoring rows or columns of a matrix is called **dominance**.

Exercise 6A

1. A two-person zero-sum game has pay-off matrix $\begin{pmatrix} 3 & -2 & 1 & -2 \\ 1 & -1 & 0 & 1 \\ -2 & 4 & 2 & -3 \end{pmatrix}$. Write down the matrix of pay-offs for player B.

2. For the following games, determine the play-safe strategies for each player and the outcomes.

   (a) $\begin{pmatrix} 5 & 4 \\ 6 & 3 \end{pmatrix}$ (b) $\begin{pmatrix} 1 & 2 \\ 4 & 3 \end{pmatrix}$ (c) $\begin{pmatrix} -4 & 0 & 2 & -2 \\ 3 & -5 & -2 & 0 \\ -3 & 1 & -5 & -2 \end{pmatrix}$ (d) $\begin{pmatrix} 4 & -3 \\ -4 & 3 \\ 0 & -4 \end{pmatrix}$

3. For each of the games in Question 2, determine if there is a stable solution. State the value of any game with a stable solution.

4 Determine if the following zero-sum games have stable solutions.

(a) $\begin{pmatrix} 4 & 5 \\ 7 & 6 \end{pmatrix}$ (b) $\begin{pmatrix} 5 & -4 \\ -5 & 4 \end{pmatrix}$ (c) $\begin{pmatrix} -4 & -3 & -2 & -1 \\ 3 & 2 & 1 & 0 \\ 2 & 1 & 0 & -1 \\ -5 & -4 & -3 & -2 \end{pmatrix}$ (d) $\begin{pmatrix} 0 & 2 \\ 4 & 6 \\ 8 & 10 \end{pmatrix}$

5 Determine play-safe strategies for the game $\begin{pmatrix} -3 & -6 & 2 & -12 \\ 12 & -11 & -5 & 7 \\ 4 & 7 & 3 & 6 \end{pmatrix}$. Can either player improve their pay-off by changing from this strategy assuming that the other still plays safe?

6 Two teams have to decide on their strategies for a cup final. Neither knows what the other will do. The probabilities of United winning with each combination of tactics is as shown in the figure.

		City		
		X	Y	Z
	1	0.2	0.6	0.1
United	2	0.3	0.5	0.6
	3	0.1	0.4	0.4

(a) Explain why United should never adopt strategy number 3.

(b) Which strategy should City never use?

(c) Reduce the table to a 2×2 pay-off matrix. Find the play-safe strategies for each team.

(d) Which strategy would you advise United to play? Explain your answer.

6.4 Mixed strategies

You can only be certain that a play-safe strategy is the best for a player when a game has a stable solution. Fortunately, there is a technique which allows you to find optimal strategies for all two-person zero-sum games. This technique uses the idea of a **mixed strategy** in which, rather than settle on a single option, a player has some likelihood of choosing any of the options. To illustrate the idea of this method, consider the game with the pay-off matrix shown in Fig. 6.7, and suppose that the game is played repeatedly.

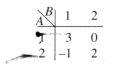

Fig. 6.7

Suppose A adopts a mixed strategy of sometimes choosing option 1 and sometimes choosing option 2. Let p be the probability of choosing option 1; the probability of choosing option 2 is then $1-p$.

If B chooses option 1

A gains 3 with probability p and gains -1 with probability $1-p$. The expected pay-off is therefore

$$3p + (-1)(1-p) = 3p - 1 + p = 4p - 1.$$

You can represent this graphically, as shown in Fig. 6.8.

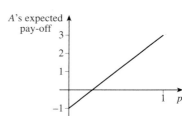

Fig. 6.8

If B chooses option 2

A gains 0 with probability p and gains 2 with probability $1-p$. The expected pay-off is therefore

$$0p + 2(1-p) = 2 - 2p.$$

You can represent this on the same diagram as Fig. 6.8, now shown as Fig. 6.9.

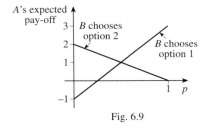

Fig. 6.9

The two lines in Fig. 6.9 cross when $4p - 1 = 2 - 2p$, that is when $p = \frac{1}{2}$. For any value of p, A can expect a pay-off that is at least the value given by the solid line in Fig. 6.10.

If A chooses a value of p less than $\frac{1}{2}$, then B's option 1 gives an expected pay-off less than 1, and B's option 2 gives an expected pay-off greater than 1. For example, if $p = \frac{1}{3}$ the expected pay-off for A is $4 \times \frac{1}{3} - 1 = \frac{1}{3}$ if B chooses option 1. The expected pay-off is $2 - 2 \times \frac{1}{3} = \frac{4}{3}$ if B chooses option 2. So A can expect to receive at least $\frac{1}{3}$.

Similarly, if A chooses a value of p greater than $\frac{1}{2}$, then B's option 2 gives a pay-off less than 1, but B's option 1 gives a pay-off greater than 1.

However, if $p = \frac{1}{2}$ then the expected pay-off will be 1 whatever B chooses. A can therefore guarantee an expected pay-off of 1 by choosing each option with probability $\frac{1}{2}$. This is illustrated by the peak in the solid line in Fig. 6.10. This expected pay-off of 1 is the greatest that A can guarantee.

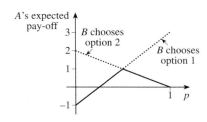

Fig. 6.10

The method that you have just used to find the maximum expected pay-off should remind you of linear programming ideas in D1 Chapter 8.

The solid line in Fig. 6.11 can be considered to be part of the boundary of a convex feasible region.

Across this region you can slide a horizontal line. Feasible points on this line all represent the same expected pay-off. When this line touches the highest vertex, the maximum expected pay-off has been reached.

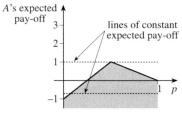

Fig. 6.11

You can carry out a similar analysis for a mixed strategy adopted by B. Let B choose options 1 and 2 with probabilities q and $1 - q$ respectively. Then the expected pay-offs to B are

If A chooses option 1   B loses $3q + 0(1 - q) = 3q$;

If A chooses option 2   B loses $(-1)q + 2(1 - q) = -q + 2 - 2q = 2 - 3q$.

The values of $3q$ and $2 - 3q$ are equal when $3q = 2 - 3q$, that is $q = \frac{1}{3}$. For this value of q, the expected pay-off is then 1 whatever A does.

The following should now be clear. For the game $\begin{pmatrix} 3 & 0 \\ -1 & 2 \end{pmatrix}$,

- A should choose options 1 and 2 with probabilities $\frac{1}{2}$ and $\frac{1}{2}$ respectively,
- B should choose options 1 and 2 with probabilities $\frac{1}{3}$ and $\frac{2}{3}$ respectively.

If both play this way, then A can expect to gain 1 and B to lose 1. The value of the game is 1.

Example 6.4.1
Determine optimal strategies for the game with pay-off matrix $\begin{pmatrix} 6 & -2 \\ -3 & -1 \end{pmatrix}$, and draw a graph to illustrate this from A's point of view.

Let A choose options 1 and 2 with probabilities p and $1-p$ respectively.

Then the expected pay-offs under B's two choices are $6p + (-3)(1-p) = 9p - 3$ and $(-2)p + (-1)(1-p) = -p - 1$.

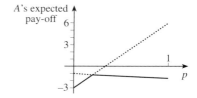

These pay-offs are equal when $9p - 3 = -p - 1$, which gives $p = \frac{1}{5}$.

The expected pay-off is then $9 \times \frac{1}{5} - 3 = -1\frac{1}{5}$.

Now that you know the game's value of $-1\frac{1}{5}$, you can use it to determine B's strategy. Let B choose options 1 and 2 with probabilities q and $1-q$ respectively.

Then the expected pay-off to B when A chooses option 1 is

$$-6q - (-2)(1-q) = 2 - 8q = 1\frac{1}{5},$$

giving $q = \frac{1}{10}$.

A should choose option 1 with probability $\frac{1}{5}$ and B should choose option 1 with probability $\frac{1}{10}$.

You could also work out B's strategy from the alternative equation derived from A choosing option 2. Thus $(-3)q + (-1)(1-q) = -1\frac{1}{5}$. This also gives $q = \frac{1}{10}$.

In practice, both players should choose their options using a randomising device with the correct probabilities.

6.5 2 × n games

The graphical method of Section 6.4 was introduced with **2 × 2 games**, that is games with pay-off matrices which have 2 rows and 2 columns. The method can be extended to $2 \times n$ games, that is games with pay-off matrices which have 2 rows and n columns.

Consider, for example, the game with the pay-off matrix shown in Fig. 6.12.

Let A choose options 1 and 2 with probabilities p and $1-p$, respectively. Then the expected pay-offs under B's choices can be represented by the straight line graphs in Fig. 6.13.

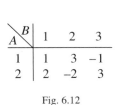

Fig. 6.12

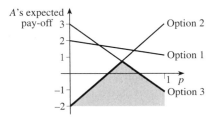

Fig. 6.13

Whatever B's strategy, A can guarantee expected pay-offs in the shaded region bounded by the thick line in Fig. 6.13.

From Fig. 6.13, A should therefore choose option 1 with probability p, determined by the point of intersection of the lines corresponding to B's option 2 and option 3.

This point of intersection is given by

$$3p + (-2)(1-p) = (-1)p + 3(1-p), \quad \text{which gives} \quad p = \tfrac{5}{9}.$$

The game thus has a value V given by

$$V = 3 \times \tfrac{5}{9} + (-2) \times \left(1 - \tfrac{5}{9}\right) = \tfrac{15}{9} - \tfrac{8}{9} = \tfrac{7}{9}.$$

B should ignore option 1 and choose option 2 with probability q, where

$$3q + (-1)(1-q) = \tfrac{7}{9}, \quad \text{or} \quad (-2)q + 3(1-q) = \tfrac{7}{9}.$$

Either way, $q = \tfrac{4}{9}$.

A general procedure for solving any $2 \times n$ game between the first player A and the second player B which does not have a stable solution is as follows.

To solve a $2 \times n$ game between A and B,

Step 1 Assign probabilities p and $1-p$ to the two options for A.

Step 2 Plot A's expected pay-off under each of B's options as lines on a graph of pay-off against p.

Step 3 Shade the region which lies underneath *every* line.

Step 4 Find the value of p at the highest point of the shaded region by solving simultaneous equations.

Step 5 Use this value of p to find the value, V, of the game.

Step 6 Assign probability q to one of those options for B which determine the highest point of the shaded region.

Step 7 Find the value of q which gives B an expected pay-off of $-V$, under either of A's options.

Example 6.5.1
Determine the optimal mixed strategies and the value for the game with the given pay-off matrix.

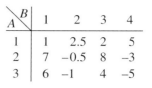

Row 3 is dominated by row 2 and so can be ignored. Column 3 is now dominated by column 1 and can also be ignored. The matrix is thus reduced to $2 \times n$ and the method given in the shaded box is appropriate. The new matrix is shown on the right.

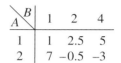

Step 1   Let A choose option 1 with probability p.

Step 2   If B plays option 1, A gains $p + 7(1-p) = 7 - 6p$.

If B plays option 2, A gains $2.5p + (-0.5)(1-p) = 3p - 0.5$.

If B plays option 4, A gains $5p + (-3)(1-p) = 8p - 3$.

You can then plot these pay-off functions.

Step 3   Shade the region that lies under every pay-off line.

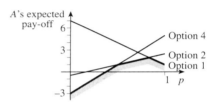

As in linear programming problems, it is easier not to shade the whole area, but simply to outline the area by shading.

Step 4   The highest point of the shaded region is where the pay-off lines for options 1 and 2 intersect, which is where

$$7 - 6p = 3p - 0.5, \quad \text{that is where} \quad p = \tfrac{5}{6}.$$

Step 5   Let V be the value of the game.

Then $V = 1 \times \tfrac{5}{6} + 7\left(1 - \tfrac{5}{6}\right) = \tfrac{5}{6} + \tfrac{7}{6} = 2$.

Step 6   The two pay-off lines which give the value of p are those for options 1 and 2. Let q be the probability that B plays option 1, so $1-q$ is the probability that B plays option 2.

Step 7   Then $1q + 2.5(1-q) = 2$, giving $q = \tfrac{1}{3}$.

The game has value 2. A should choose options 1 and 2 with probabilities $\tfrac{5}{6}$ and $\tfrac{1}{6}$ respectively. B should choose options 1 and 2 with probabilities $\tfrac{1}{3}$ and $\tfrac{2}{3}$ respectively.

An $n \times 2$ game can be transformed into a $2 \times n$ game by interchanging the two players. In order that the pay-offs continue to relate to the player whose choices are listed at the left, the signs of all the entries in the matrix should be changed.

Example 6.5.2
Find the value of the game with the pay-off matrix $\begin{pmatrix} -2 & 0 \\ 1 & -2 \\ -3 & 2 \end{pmatrix}$.

Consider instead the game with matrix $\begin{pmatrix} 2 & -1 & 3 \\ 0 & 2 & -2 \end{pmatrix}$, where the rows and columns have been interchanged, and the signs of the entries have been changed.

Let A choose option 1 with probability p.

If B plays option 1, A gains $2p + 0(1-p) = 2p$.

If B plays option 2, A gains
$(-1)p + 2(1-p) = 2 - 3p$.

If B plays option 3, A gains
$3p + (-2)(1-p) = 5p - 2$.

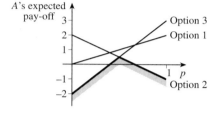

You can find the value of p by solving the equation arising from equating the pay-offs of options 2 and 3.

Thus $2 - 3p = 5p - 2$, giving $p = \tfrac{1}{2}$.

The value $V = (-1) \times \tfrac{1}{2} + 2(1 - \tfrac{1}{2}) = \tfrac{1}{2}$.

Recalling that the signs of the original game were changed, the value of the original game is therefore $-\tfrac{1}{2}$.

Exercise 6B

1 A two-person zero-sum game has the pay-off matrix shown in the figure.

Suppose that Ann uses random numbers to choose strategy M with probability p.

	Ben		
Ann	X	Y	Z
M	3	-2	1
N	1	3	2

(a) Find the expected gains when Ben chooses each of strategies X, Y and Z.

(b) Use a graphical method to find the optimum value for p.

(c) State the expected value of the game for both Ann and Ben.

(d) Which strategy should Ben never use?

2 A two-person game has pay-off matrix shown in the figure.

	X	Y
A	5	2
B	3	4

(a) Suppose that the first player chooses strategy A with probability p. Find the expected gains for the first player when the second player chooses each of strategies X and Y.

(b) Suppose that the second player chooses strategy X with probability q. Find the expected gains for the first player when the first player chooses each of strategies A and B.

(c) Find the value of the game and the corresponding values of p and q.

3 Find the values of the following games.

(a) $\begin{pmatrix} 4 & 1 \\ 3 & 5 \end{pmatrix}$
(b) $\begin{pmatrix} 2 & 1 \\ 1 & 3 \end{pmatrix}$
(c) $\begin{pmatrix} -1 & 3 \\ 2 & -1 \end{pmatrix}$

4 (a) Use dominance to reduce the game with pay-off matrix $\begin{pmatrix} 6 & 7 & 2 \\ 1 & 3 & 5 \end{pmatrix}$ to a 2×2 game.

(b) Hence find the value of this game.

5 The Rolling Pebbles have been playing as a band for many years. When they tour they sometimes play old songs, they sometimes play new songs and they sometimes play a mixture of old and new songs. The table shows the audience reaction (as a score out of 10) for each of the possible combinations. High scores are good.

		Audience		
		Young	Mixed	Older
Songs played	Old	1	3	8
	Mixture	3	1	2
	New	8	5	3

Explain why, according to these data, the band should never choose to play a mixture of old and new songs.

The band do not know whether their audience will be young, older or of mixed ages. Suppose that they choose to play old songs with probability p and new songs with probability $1 - p$.

(a) Calculate, in terms of p, the expected reaction from each of the three types of audience.

(b) Use a graphical method to decide what value p should take to maximise the minimum expected reaction from part (a). Mark clearly on your graph the vertex where the optimal value occurs.

(OCR)

6.6 $m \times n$ games

The graphical method of Section 6.5 cannot be applied when the pay-off matrix has more than 2 rows. Fortunately, however, it is possible to convert the ideas of Section 6.5 into linear programming form. This method will be illustrated with a 3×2 game but can be applied to a game of any size.

Consider the game with pay-off matrix $\begin{pmatrix} 2 & 4 \\ 5 & 2 \\ 1 & 6 \end{pmatrix}$.

Suppose the first player, A, chooses options 1, 2 and 3 with probabilities p_1, p_2 and p_3 respectively. Then the expected pay-offs are $2p_1 + 5p_2 + 1p_3$ and $4p_1 + 2p_2 + 6p_3$, depending upon which strategy the second player chooses. Let v be the pay-off which the first player is trying to maximise. Then, in linear programming format, the problem is maximise v,

subject to $\left.\begin{array}{l}v \leqslant 2p_1 + 5p_2 + 1p_3, \\ v \leqslant 4p_1 + 2p_2 + 6p_3,\end{array}\right\}$

$p_1 + p_2 + p_3 \leqslant 1$,

$p_1, p_2, p_3 \geqslant 0$.

These two conditions ensure that v is the smaller of $2p_1 + 5p_2 + 1p_3$ and $4p_1 + 2p_2 + 6p_3$.

Note the use of the condition $p_1 + p_2 + p_3 \leqslant 1$. Because all the elements in the pay-off matrix are positive, v is bound to be positive. When v is maximised, $p_1 + p_2 + p_3$ will therefore automatically equal 1, as required, because they are all probabilities.

The problem can now be solved in the standard manner, using the Simplex method given in Chapter 5. Equation numbers will be written in bold type. Here is the tableau for the problem.

P	v	p_1	p_2	p_3	r	s	t		Equation
1	-1	0	0	0	0	0	0	0	**1**
0	0	1	1	1	1	0	0	1	**2**
0	1	-2	-5	-1	0	1	0	0	**3**
0	1	-4	-2	-6	0	0	1	0	**4**
1	0	-4	-2	-6	0	0	1	0	**5** = **1** + **8**
0	0	1	1	1	1	0	0	1	**6** = **2**
0	0	2	-3	5	0	1	-1	0	**7** = **3** − **8**
0	1	-4	-2	-6	0	0	1	0	**8** = **4**
1	0	0	-8	4	0	2	-1	0	**9** = **5** + 4 × **11**
0	0	0	$\frac{5}{2}$	$-\frac{3}{2}$	1	$-\frac{1}{2}$	$\frac{1}{2}$	1	**10** = **6** − **11**
0	0	1	$-\frac{3}{2}$	$\frac{5}{2}$	0	$\frac{1}{2}$	$-\frac{1}{2}$	0	**11** = $\frac{1}{2}$ × **7**
0	1	0	-8	4	0	2	-1	0	**12** = **8** + 4 × **11**
1	0	0	0	$-\frac{4}{5}$	$\frac{16}{5}$	$\frac{2}{5}$	$\frac{3}{5}$	$\frac{16}{5}$	**13** = **9** + 8 × **14**
0	0	0	1	$-\frac{3}{5}$	$\frac{2}{5}$	$-\frac{1}{5}$	$\frac{1}{5}$	$\frac{2}{5}$	**14** = $\frac{2}{5}$ × **10**
0	0	1	0	$\frac{8}{5}$	$\frac{3}{5}$	$\frac{1}{5}$	$-\frac{1}{5}$	$\frac{3}{5}$	**15** = **11** + $\frac{3}{2}$ × **14**
0	1	0	0	$-\frac{4}{5}$	$\frac{16}{5}$	$\frac{2}{5}$	$\frac{3}{5}$	$\frac{16}{5}$	**16** = **12** + 8 × **14**
1	0	$\frac{1}{2}$	0	0	$\frac{7}{2}$	$\frac{1}{2}$	$\frac{1}{2}$	$\frac{7}{2}$	**17** = **13** + $\frac{4}{5}$ × **19**
0	0	$\frac{3}{8}$	1	0	$\frac{5}{8}$	$-\frac{1}{8}$	$\frac{1}{8}$	$\frac{5}{8}$	**18** = **14** + $\frac{3}{5}$ × **19**
0	0	$\frac{5}{8}$	0	1	$\frac{3}{8}$	$\frac{1}{8}$	$-\frac{1}{8}$	$\frac{3}{8}$	**19** = $\frac{5}{8}$ × **15**
0	1	$\frac{1}{2}$	0	0	$\frac{7}{2}$	$\frac{1}{2}$	$\frac{1}{2}$	$\frac{7}{2}$	**20** = **16** + $\frac{4}{5}$ × **19**

The game has value $\frac{7}{2}$. The first player should choose $p_1 = 0$, $p_2 = \frac{5}{8}$, $p_3 = \frac{3}{8}$.

The method just employed requires the elements of the pay-off matrix to be positive (to ensure that the sum of the probabilities equals 1). If this is not the case, then you can simply add the same number to every element, remembering that you have then increased the value by this number.

CHAPTER 6: GAME THEORY

You will have seen that, although linear programming using the Simplex method allows you to solve problems of any size, in practice the working is very time consuming. For that reason, most of the problems you will consider in this section will be of a relatively small size.

Example 6.6.1

Convert the problem of Example 6.4.1, with pay-off matrix $\begin{pmatrix} 6 & -2 \\ -3 & -1 \end{pmatrix}$ into a linear programming problem. Find the optimum strategy for the first player by using the Simplex algorithm.

First, add 4 to every element, so that the pay-off matrix is now $\begin{pmatrix} 10 & 2 \\ 1 & 3 \end{pmatrix}$.

The problem is

$$\begin{aligned} \text{maximise} \quad & v - 4, \\ \text{subject to} \quad & v \leq 10p_1 + p_2, \\ & v \leq 2p_1 + 3p_2, \\ & p_1 + p_2 \leq 1, \\ & p_1, p_2 \geq 0. \end{aligned}$$

P	v	p_1	p_2	r	s	t		Equation
1	−1	0	0	0	0	0	−4	**1**
0	1	−10	−1	1	0	0	0	**2**
0	1	−2	−3	0	1	0	0	**3**
0	0	1	1	0	0	1	1	**4**
1	0	−2	−3	0	1	0	−4	**5 = 1 + 7**
0	0	−8	2	1	−1	0	0	**6 = 2 − 7**
0	1	−2	−3	0	1	0	0	**7 = 3**
0	0	1	1	0	0	1	1	**8 = 4**
1	0	−14	0	$\frac{3}{2}$	$-\frac{1}{2}$	0	−4	**9 = 5 + 3 × 10**
0	0	−4	1	$\frac{1}{2}$	$-\frac{1}{2}$	0	0	**10 = $\frac{1}{2}$ × 6**
0	1	−14	0	$\frac{3}{2}$	$-\frac{1}{2}$	0	0	**11 = 7 + 3 × 10**
0	0	5	0	$-\frac{1}{2}$	$\frac{1}{2}$	1	1	**12 = 8 − 10**
1	0	0	0	$\frac{1}{10}$	$\frac{9}{10}$	$\frac{14}{5}$	$-\frac{6}{5}$	**13 = 9 + 14 × 16**
0	0	0	1	$\frac{1}{10}$	$-\frac{1}{10}$	$\frac{4}{5}$	$\frac{4}{5}$	**14 = 10 + 4 × 16**
0	1	0	0	$\frac{1}{10}$	$\frac{9}{10}$	$\frac{14}{5}$	$\frac{14}{5}$	**15 = 11 + 14 × 16**
0	0	1	0	$-\frac{1}{10}$	$\frac{1}{10}$	$\frac{1}{5}$	$\frac{1}{5}$	**16 = $\frac{1}{5}$ × 12**

The game has value $-\frac{6}{5}$. The first player should play options 1 and 2 with probabilities $\frac{1}{5}$ and $\frac{4}{5}$, respectively.

The second player's strategy can be found by a similar procedure or, more simply, by solving $6q - 2(1-q) = -\frac{6}{5}$, giving $q = \frac{1}{10}$.

Exercise 6C

1 A two-person zero-sum game has pay-off matrix $\begin{pmatrix} 4 & 1 \\ 3 & 5 \end{pmatrix}$.

Suppose the first player chooses options 1 and 2 with probabilities p and q respectively. Let v be the expected pay-off.

Formulate this as a linear programming problem and hence find v.

2 A two-person zero-sum game has pay-off matrix $\begin{pmatrix} -1 & 3 \\ 2 & -1 \end{pmatrix}$.

By first adding 2 to every element, convert the game into a linear programming problem and hence find the value of the game.

3 For the game with pay-off matrix $\begin{pmatrix} 3 & 2 & 5 \\ 4 & 5 & 2 \\ 2 & 4 & 4 \end{pmatrix}$, suppose the first player chooses options 1, 2 and 3 with probabilities p, q and r respectively. Formulate, but do not solve, the problem of finding the game's value v as a linear programming problem.

Miscellaneous exercise 6

1 A game has pay-off matrix shown in the figure.

Show that this game has a stable solution and find the play-safe strategies for each player.

A\B	1	2	3	4
1	6	−3	15	−11
2	7	1	9	5
3	−3	0	−5	8

2 The game 'stone-scissors-paper' has pay-off matrix shown.

	St	Sc	P
St	0	1	−1
Sc	−1	0	1
P	1	−1	0

(a) Suppose that the first player chooses stone, scissors and paper with probabilities p, q and $1-p-q$ respectively. Find the expected gains when the second player chooses each of the strategies stone, scissors and paper.

(b) How can the first player guarantee an expected return of 0?

(c) What is the value of the game? Justify your answer.

3 The pay-off matrix for a zero-sum game between two players A and B is $\begin{matrix} & B \\ A & \begin{pmatrix} 3 & -2 \\ -2 & 1 \end{pmatrix} \end{matrix}$.

(a) Show that the game does not have a stable solution.

Player A uses the mixed strategy defined by $(0.6, 0.4)$.

(b) Determine the expected pay-off for A if player B:

    (i) plays column one;

    (ii) plays column two;

    (iii) adopts the strategy $(0.5, 0.5)$.

(c) Determine the optimal strategy for A and its expected pay-off. (OCR)

CHAPTER 6: GAME THEORY

4 Roland and Colleen play a two-person zero-sum game. The table shows the pay-off matrix for the game. The values in the table are the amounts won by Roland.

		Colleen	
		Stick	Twist
Roland	Stick	−1	4
	Twist	3	−2

(a) Find Roland's and Colleen's play-safe strategies, and hence show that this game does not have a stable solution.

(b) (i) State which strategy Roland should choose if he knows that Colleen will choose her play-safe strategy.

(ii) State which strategy Colleen should choose if she knows that Roland will choose his play-safe strategy.

Roland and Colleen play the game a large number of times. Colleen uses random numbers to choose the Stick strategy with probability p.

(c) Show that the expected gain for Roland when he chooses the Stick strategy is given by $4 - 5p$, and find a similar expression for the expected gain for Roland when he chooses the Twist strategy.

(d) Use a graphical method to find the optimum value of p. (OCR)

5 Robin is playing a computer game in which he has to protect the environment. He chooses an energy source and the computer chooses the weather conditions.

The numbers of points scored by Robin under each of the combinations of energy type and weather conditions are shown in the table.

		Computer		
		Warm	Wet	Windy
	Atomic energy	−6	3	5
Robin	Bio-gas	2	4	6
	Coal	5	1	3

Robin is trying to maximise his points total and the computer tries to stop him.

(a) Explain why Robin should not choose Atomic energy and why the computer should not choose Windy weather.

(b) Find the play-safe strategies for the reduced game for Robin and for the computer, and hence show that this game does not have a stable solution.

Suppose that Robin uses random numbers to choose Bio-gas with probability p and Coal with probability $1 - p$.

(c) Show that the expected loss for the computer when it chooses Warm weather is given by $5 - 3p$, and find an expression for the expected loss when it chooses Wet weather.

(d) Use a graphical method to find the optimum value of p and the corresponding expected gain for Robin. (OCR)

6 Roy and Callum play a two-person zero-sum game. The table shows the pay-off matrix for the game. The values in the table are the amounts won by Roy.

		Callum	
		Strategy A	Strategy B
Roy	Strategy P	−1	1
	Strategy Q	4	−3

(a) Find Roy's and Callum's play-safe strategies, and show that this game does not have a stable solution.

(b) (i) State which strategy Roy should choose if he knows that Callum will always choose his play-safe strategy.

(ii) State which strategy Callum should choose if he knows that Roy will always choose his play-safe strategy.

Suppose that Roy uses random numbers to choose strategy P with probability p.

(c) Show that the expected gain for Callum when he chooses strategy A is given by $5p - 4$, and find a similar expression for the expected gain for Callum when he chooses strategy B.

(d) Use a graphical method to find the optimum value of p and the corresponding expected gain for Roy. (OCR)

7 Rowena and Colin play a two-person zero-sum simultaneous-play game. The table shows the pay-off matrix for the game.

		Colin		
		Strategy X	Strategy Y	Strategy Z
Rowena	Strategy A	4	−1	2
	Strategy B	4	6	3
	Strategy C	1	2	−2

(a) Find Rowena's and Colin's play-safe strategies, and hence show that this game has a stable solution.

(b) Explain what having a stable solution means to the way the game is played.

(c) Explain why Colin will never choose strategy X, and hence reduce the game to give a 2×2 pay-off matrix.

Suppose that Colin uses random numbers to choose strategy Y with probability p.

(d) Show that the expected gain when Rowena chooses strategy A is given by $2 - 3p$ and find a similar expression for the expected gain when Rowena chooses her other strategy.

(e) Use a graphical method to find the optimum value of p and the corresponding expected gain for Colin. (OCR)

8 Richard and Carol play a two-person zero-sum simultaneous-play game. The table shows the pay-off matrix for the game.

		Carol		
		Strategy X	Strategy Y	Strategy Z
Richard	Strategy A	2	3	−2
	Strategy B	−4	−1	−1
	Strategy C	−5	0	1

(a) Explain the meaning of the term zero-sum game.

(b) Find the play-safe strategies for both Richard and Carol, and hence show that this game does not have a stable solution.

(c) Suppose that Richard knows that Carol will use her play-safe strategy. Explain whether or not he should change from his play-safe strategy as found in part (b).

(d) Suppose that Carol knows that Richard will use his play-safe strategy. Explain whether or not she should change from her play-safe strategy as found in part (b). (OCR)

9 Rose is playing a computer game in which she has to defend a planet from aliens. She chooses a defence strategy and the computer chooses an attack strategy.

The number of points scored by Rose with each combination of strategies is shown in the table.

		Computer		
		Fight	Shoot	Track
Rose	Delay	−2	−5	1
	Hide	3	4	6
	Negotiate	5	−1	2

Rose is trying to maximise the number of points that she scores, and the computer is trying to minimise the number of points that Rose scores.

(a) Find the play-safe strategy for Rose and for the computer, and hence show that this game does not have a stable solution.

(b) Explain why Rose will not choose the Delay strategy.

(c) Which strategy will the computer never choose to play?

Suppose that Rose uses random numbers to choose between her two remaining strategies, choosing the Hide strategy with probability p and the Negotiate strategy with probability $1-p$.

(d) Find expressions for the expected gain for Rose when the computer chooses each of its remaining strategies.

(e) Calculate the value of p for Rose to maximise her guaranteed return. (OCR)

10 Richard is playing a computer battle game in which he chooses a warrior and the computer chooses an opponent. Neither Richard nor the computer knows what the other has chosen.

The probabilities of Richard winning the battle with each combination of warrior and opponent are shown in the table. If Richard does not win, the computer wins. Both Richard and the computer are playing to win.

		Computer		
		Dragon	Elf	Fighter
Richard	Argent	0.4	0.7	0.3
	Bronze	0.5	0.6	0.8
	Crystal	0.2	0.5	0.1

(a) Explain why the computer should never choose Elf.

(b) Which warrior should Richard never choose?

(c) Use the information from parts (a) and (b) to reduce the table to a 2×2 matrix. Find the play-safe strategies for the reduced game for Richard and for the computer.

(d) Richard plays the game many times. What strategy should he use? Explain your answer.

(OCR)

11 A stockbroker, B, has three strategies to tempt clients from its main competitor, C, which are referred to as X, Y and Z for simplicity. The competing stockbroker, C, also has three marketing strategies which are referred to as P, Q and R. The table below shows the result, from previous experience, of these strategies from B's point of view (that is, if B decides upon strategy X and C chooses strategy P, then B will lose 2 customers to C). Each stockbroker can only adopt a single marketing strategy in any one month.

		Stockbroker C		
		P	Q	R
Stockbroker B	X	−2	5	−3
	Y	3	−5	−2
	Z	1	8	−1

(a) Show that, regarding this as a zero-sum game, there is no stable solution.

(b) Show that it will never be optimal for B to adopt strategy X.

(c) By considering mixed strategies, find the optimal mixed marketing strategy for B, giving any probability as an exact fraction.

Revision exercise

1 As part of their Leisure and Tourism course, some students are planning to produce a video to promote the attractions of their town. Some of the activities involved are listed below.

	Activity	Expected duration (days)	Preceded by
A	Visit possible locations	1	—
B	Plan storyboard and write script	2	—
C	Plan shooting schedule	1	A, B
D	Get permission at locations	1	C
E	Organise lighting, sound and equipment	2	C
F	Organise performers	1	C
G	Outdoor filming	2	D, E, F
H	Indoor filming	3	E, F
I	Editing and sound dubbing	1	G, H

(a) Copy and complete the activity schedule below, showing the earliest and latest start and finish times of each activity (assuming completion in the minimum possible time), and the float for each activity.

Activity	Earliest start	Latest start	Earliest finish	Latest finish	Float
A	0	1	1	2	1
B	0	0	2	2	0
C					
D					
E					
F	3	4	4	5	1
G					
H					
I					

(b) Find the minimum time for completion of the project.

(c) List the critical activities.

(d) Suppose that poor weather holds up the outdoor filming (activity G) by 1 day. Describe the effect that this will have on the total time to complete the project. (OCR)

2   The Rolling Pebbles have been invited to appear on three different talk shows. The shows are all at the same time, but in different places. The four members of the band have decided that one of them will not appear on any show, and that each of the other three will appear on a different show.

The scores in the table show how popular each band member is likely to be on each show. A higher score means that the band member is more popular. The band wants to find the allocation that gives the maximum total score.

		Band member			
		Wart	Xenon	Yogi	Ziggy
Show	Local radio	5	6	3	4
	Midday TV	3	7	4	2
	National radio	7	8	6	4

(a) Explain how the figures in the table can be changed so that the problem becomes one of finding an allocation that gives a minimum non-negative total.

(b) Explain how to convert the adapted scores from part (a) into a square matrix on which the Hungarian algorithm can be used.

(c) Use the Hungarian algorithm, reducing rows first, to allocate the band members to the shows in the most appropriate way. (OCR)

3   The diagram on the left shows an undirected network. S is the source and T is the sink. The values show the capacities of the arcs.

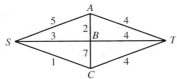 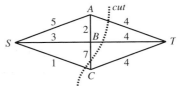

(a) Redraw the left diagram as a directed flow network.

A *cut* can be drawn through the undirected network, as shown in the diagram on the right.

(b) Define what is meant by a cut.

A cut can be described by listing the vertices which are on the source side, X, and the vertices which are on the sink side, Y. The cut shown has $X = \{S, A, B\}$ and $Y = \{C, T\}$.

(c) Write down the number of different cuts that are possible for this network (including the cut shown).

(d) List all the cuts from part (c), by giving the vertices in X and the vertices in Y, and calculate the value of each cut.

(e) Explaining your reasoning carefully, describe what you can conclude from the results of part (d).

4 Visitors to a stately home tour the house on one of a number of routes. The diagram shows these routes and the maximum flows in people per hour.

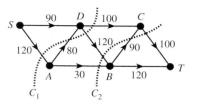

(a) List the eight possible routes from S to T.

(b) Calculate the values of the cuts C_1 and C_2, and explain what this tells you about the maximum flow from S to T.

(c) Find the maximum flow from S to T.

5 The activities involved in preparing a major archaeological dig are listed in the table below.

	Activity	Expected duration (days)	Number of people needed	Preceded by
A	Survey site	2	3	–
B	Aerial photography	2	2	A
C	Geophysics	5	2	A
D	Remove topsoil	2	3	A
E	Check records of previous digs	1	2	A
F	Dig investigation trench	4	5	D, E
G	Check old maps and public records	5	2	E
H	Computer analysis	3	2	B, C
I	Begin main excavation	10	5	F, G
J	Record findings	4	4	H, I

(a) Draw up a table showing the earliest and latest possible start and finish times for each activity (in days from the beginning of the project).

(b) Write down the critical activities.

The table below shows the number of people needed each day when using the earliest starting date schedule. This schedule requires a minimum of nine people in the team.

Day number	1	2	3	4	5	6	7	8	9	10	11
Team size	3	3	9	9	9	9	9	9	7	7	5
Day number	12	13	14	15	16	17	18	19	20	21	22
Team size	5	5	5	5	5	5	5	4	4	4	4

(c) Draw up a schedule that is consistent with the earliest finish time but only needs seven people in the team.

(OCR)

6 A relay team consists of four runners, A, B, C and D, each of whom runs one leg of the race. The best training times, in seconds, for each of the runners over each of the legs are given in the table.

	1st leg	2nd leg	3rd leg	4th leg
A	47	45	45	43
B	48	44	45	44
C	46	45	44	43
D	50	47	46	44

Use the Hungarian algorithm to decide which runner should be allocated to which leg of the race.

7 The diagram shows a network with (stage, state) variables at the vertices and costs on the edges.

Set up and solve a dynamic programming tabulation to find the route from $(3,1)$ to $(0,1)$ for which the costs are minimum.

8 (a) Display the linear programming problem (on the right) in a Simplex tableau.

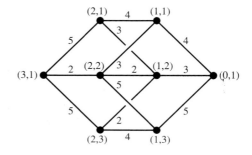

Maximise $P = 4x + 2y + 3z$,

subject to $3x + 7y + 2z \leq 15$,
$2x + 4y + z \leq 8$,

and $x \geq 0, y \geq 0, z \geq 0$.

(b) Use the Simplex method to perform one iteration of your tableau for part (a), choosing to pivot on an element chosen from the x-column, writing the entry as a fraction where appropriate.

(c) State the values of x, y, z and P resulting from this iteration. State whether or not this is the optimal solution and give a reason for your answer. (AQA)

9 Rachael is playing a two-person zero-sum simultaneous-play game against her computer. The pay-off matrix is as shown.

		Computer		
		2	3	−2
Rachael		−4	−1	−1
		−5	0	1

Rachael decides to choose her options with probabilities p_1, p_2 and p_3, respectively, and wants to maximise the pay-off.

(a) By first adding 6 to every element in the pay-off matrix, formulate this as a linear programming problem.

(b) Show (but do not solve) the initial Simplex tableau.

10 Minnie, Max and their friend Dominic have formed a quiz team. Much to their surprise they have reached the final of a mathematics quiz. They are required to nominate one team member to take part in a solo round.

The solo round will consist of 10 questions on one of three specialist subjects: Arabic, Babylonian or Chinese mathematics. The contestants do not know which of the specialist subjects has been chosen.

Based on their performances in rehearsal, the members of the team expect their scores out of 10 to be as given in the table below.

		Specialist subject		
		Arabic	Babylonian	Chinese
Solo player	Minnie	4	7	4
	Max	6	5	7
	Dominic	3	6	3

(a) Explain why Dominic should not be chosen to play the solo round.

(b) Dominic suggests tossing a coin and letting Minnie play the solo round if the coin comes up heads, and Max play if it comes up tails. Assuming that the coin is fair, calculate the expected score for each of the three possible specialist subjects.

(c) If, instead of using equal probabilities, they choose Minnie with probability p and Max with probability $1-p$ find the expected score for each of the three possible specialist subjects.

(d) Minnie objects that this is not a repeated play situation and suggests that the team should use their play-safe strategy. Showing your working, find out which member of the team should play the solo round if they use this strategy. (OCR)

11 The network represents a group of cities through which a haulage firm could travel on the way from S to T. The weights represent the costs, in hundreds of pounds, for each possible leg.

Use dynamic programming to find the optimal route and the total cost (profit).

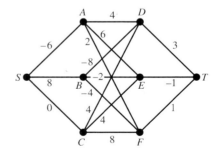

[Negative numbers represent profits.]

Mock examination 1

Time 1 hour 20 minutes

Answer all the questions.
You are permitted to use a graphic calculator in this paper.

1. The diagram shows a system of pipes and their capacities, in litres per second.

   (a) List all 8 possible cuts and calculate their capacities. [5]

   (b) Draw a diagram showing the maximum possible flow from S to T and state how you know it is a maximum. [3]

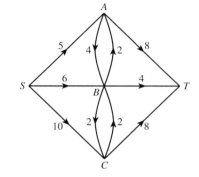

2. The diagram shows three stages in a manufacturing process. The vertices are labelled using (stage, state) and the values on the edges represent costs in £s.

   Set up a dynamic programming tabulation to find the route which gives the least total cost. [8]

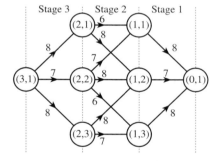

3. Consider the following linear programming problem:
   maximise $P = 3x + 2y,$

   subject to $x + y \leqslant 10,$
   $x + 2y \leqslant 15,$
   $3x + y \leqslant 22,$
   $x \geqslant 0, y \geqslant 0.$

   (a) Write down an initial Simplex tableau for this problem. Perform one iteration of the Simplex method, choosing to pivot on an element in the x-column. [5]

   (b) Complete the Simplex method to find the maximum value of P. Give the values of the slack variables at the maximum point and say what information is given about each inequality by the value of its slack variable. [5]

4 The installation of some self-assembly kitchen units has been divided into the nine activities shown below.

Activity	Duration (minutes)	Immediate predecessors
A	10	–
B	20	A
C	45	A
D	50	B
E	65	B
F	10	B
G	25	C, F
H	12	E
I	10	D, G, H

(a) Draw an activity network for this project and find the earliest starting time and latest finish time for each task. [5]

(b) Give the fastest completion time and the critical path. [2]

(c) Construct a cascade (Gantt) diagram for the project, assuming each activity is to start as early as possible. [3]

5 Five job applicants receive the following scores for their suitability for four positions in a company.

	I	II	III	IV
Amy	7	6	8	7
Brad	9	8	7	8
Cath	7	6	8	8
Des	6	6	7	7
Ed	8	8	8	9

It is intended to allocate the positions so as to maximise the total score.

(a) The Hungarian algorithm finds the allocation with the minimum total cost. Show how this problem can be converted into a minimisation problem. [2]

(b) The Hungarian algorithm requires the matrix to be square. Explain how to represent this problem as a square matrix. [2]

(c) Use the Hungarian algorithm, reducing rows first, to pair the applicants to the positions. [8]

6 Ken has a slightly quicker dinghy than Patrick and so will tend to win any race if they choose the same tactics. On a particular stretch of a race they can each choose to sail left (*L*), right (*R*) or tack repeatedly on a central path (*C*). The expected gains, in minutes, for Ken are shown in the table.

		Patrick		
		L	*C*	*R*
	L	2	0	2
Ken	*C*	3	3	1
	R	3	0	1

(a) Explain why Ken should never adopt strategy *R*. [1]

(b) Which strategy should Patrick never use? [1]

(c) Reduce the table to a 2×2 pay-off matrix. Find the play-safe strategies for each sailor and show that this game does not have a stable solution. [4]

(d) By considering mixed strategies, find the optimal strategy for Patrick. What is then the expected gain for Ken? [6]

Mock examination 2

Time 1 hour 20 minutes

Answer all the questions.
You are permitted to use a graphic calculator in this paper.

1. Consider the linear programming problem:

    maximise $P = 3x + y$,
    subject to $x + 2y \leq 8$,
    $2x + y \leq 10$,
    $x \geq 0, y \geq 0$.

    (a) Represent this problem as an initial Simplex tableau. [2]

    (b) Carry out one iteration of the Simplex algorithm, choosing to pivot on an element in the x-column. [3]

    (c) State the maximum value of P, explaining how you know that the maximum has been achieved. [2]

2. It is required to find the shortest route from G to A.

    (a) Explain why Dijkstra's algorithm is not an appropriate method for this problem. [1]

    (b) Set up and use a dynamic programming tabulation to find the shortest route from G to A. [8]

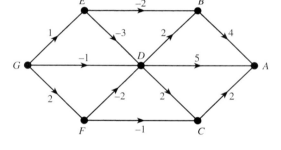

3. A company obtains quotations for the costs of placing four different advertisements with four newspapers. These costs are shown in the table.

	Daily Echo	Earth Today	News in Britain	The Galaxy
Ad A	300	540	820	300
Ad B	400	480	640	320
Ad C	380	460	540	380
Ad D	320	560	640	400

    It is decided to place one of each type of advertisement, with each newspaper being used for just one advertisement. Use the Hungarian algorithm, reducing rows first, to minimise the total cost of this policy. [9]

4 A network has vertices S, T, A, B and C. To find the maximum flow from S to T, a linear programme is set up as follows where, for example, x_{AB} is the flow along edge AB.

Maximise $\quad x_{SA} + x_{SB} + x_{SC}$
subject to $\quad x_{SA} - x_{AB} - x_{AT} = 0,$
$\quad x_{SA} + x_{AB} - x_{BC} - x_{BT} = 0,$
$\quad x_{SC} + x_{BC} - x_{CT} = 0,$
$\quad x_{SA} \leqslant 2, x_{SB} \leqslant 3, x_{SC} \leqslant 1,$
$\quad x_{AT} \leqslant 1, x_{BT} \leqslant 2, x_{CT} \leqslant 3,$
$\quad x_{AB} \leqslant 1, x_{BA} = 0,$
$\quad x_{BC} \leqslant 1, x_{CB} = 0,$
all x-values $\geqslant 0$.

(a) Explain briefly the purpose of the three equality constraints. [2]

(b) Draw the network, showing the maximum allowed flows along the arcs. [3]

(c) Explain the form taken by the objective function and give an alternative objective function. [2]

(d) For the network of part (b), find a flow and a cut having the same value. Explain what you can conclude about this value, justifying your answer. [3]

5 A company is planning the completion of a building project. The project has been divided up into the activities shown in the table.

Task	Duration (days)	Immediate predecessors	Number of workers required
A	2	–	3
B	4	A	2
C	1	A	2
D	3	B, C	4
E	2	D	2
F	3	D	2
G	4	D	3
H	3	E	2
I	3	F, G	1
J	2	H, I	2
K	6	G	2
L	1	J, K	4

(a) (i) Draw an activity network for this project. [2]

(ii) Given that the project is to be completed as soon as possible, find the earliest starting time and the latest finishing time for each task.

Give the fastest finishing time and the critical path. [5]

(b) Construct a resource histogram for the project showing the number of workers required each day, assuming that each activity is to start as early as possible. [2]

(c) Find **one** possible way that the activities can be rescheduled so that the fastest completion time is attained, but no more than 5 workers are required on any day. [2]

6 Two opposing football clubs, Newton Heath and Real Valencia, have to decide on their teams' initial line-ups for the European Cup Final. Neither knows what the other will do. The probabilities of Real Valencia gaining an initial advantage are shown in the pay-off matrix given below. Treat these probabilities as the 'gains' for Real Valencia.

		Newton Heath	
		Strategy X	Strategy Y
Real Valencia	Strategy A	0.2	0.4
	Strategy B	0.5	0.1

(a) Find the play-safe strategies for both teams, and hence show that this game does not have a stable solution. [3]

(b) Suppose that Real Valencia know that Newton Heath will use their play-safe strategy. Explain whether or not they should change from their play-safe strategy as found in part (a). [1]

(c) Suppose that Newton Heath know that Real Valencia will use their play-safe strategy. Explain whether or not Newton Heath should change from their play-safe strategy as found in part (a). [1]

Suppose that Newton Heath uses random numbers to choose strategy X with probability p.

(d) Show that the expected gain when Real Valencia chooses strategy A is given by $0.4 - 0.2p$ and find a similar expression for the expected gain when Real Valencia chooses strategy B. [3]

(e) Use a graphical method to find the optimum value of p and the corresponding expected gain for Real Valencia. [4]

(f) What strategy should Real Valencia use? [2]

Answers

1 Allocation

Exercise 1 (page 9)

1. (a) Add a column of 74s.
   (b) Back-*E*, Breast-*D*, Butterfly-*A*, Crawl-*B*
2. (a) Subtract each element from 19.
   (b) Ali-Music, Bea-Sport, Chris-Science, Deepan-Literature
3. $1+1+3+1+2+1=9$
4. 23
5. (a) $5 \times 5 \times 2 - 3 \times 5 \times 2 = 20$
   (b) $(n-k)nl$
6. *A*-1, *B*-4, *C*-3, *D*-2
7. 1-Stephen, 2-Josh, 3-Tom, 4-Haile

Miscellaneous exercise 1 (page 11)

1. (a) Jenny-Drainage, Kenny-Building, Lenny-Carpentry, Penny-Electrics
   (b) £14,000
2. (a) Add a row of 0.51s.
   (b) $\begin{pmatrix} 0 & 0.02 & 0.02 & 0 & 0.06 \\ 0.02 & 0.05 & 0 & 0.02 & 0.31 \\ 0.02 & 0 & 0.02 & 0 & 0.38 \\ 0.02 & 0 & 0.01 & 0.03 & 0 \\ 0.32 & 0.22 & 0.34 & 0 & 0.32 \end{pmatrix}$
   An allocation of 5 zeros is now possible, so this gives a minimum probability allocation.
   (c) *W-A*, *T-C*, *F-B*, *S-E*
3. (a) 1-*M*, 2-*L*, 3-*S*, 4-*T*
   (b) £119,000
4. Purple-Flowers, Orange-Diamonds, Red-Gaudy, Yellow-Circles
   or Purple-Flowers, Orange-Gaudy, Red-Circles, Yellow-Diamonds
5. (a) *W-C*, *T-A*, *F-E*, *S-B*
   (b) £16.50
6. (a) Subtract all elements from 5.
   (b) Add a row of 5s.
   (c) *A-M*, *B-V*, *C-P*, *D-S*, *E-R*
7. *B-P*, *C-Q*, *D-R*, *E-S*
8. (a) For example, rows 2, 4, 7; columns 1, 3, 4, 8
   (b) 1   (c) 2

2 Network flows

Exercise 2A (page 17)

1. 18, 20, 21, 24
   The maximum flow is *at most* 18.

2. (a) 6   (b) Cut *SB*, *AB* and *AT*.
3. (a) 35 000   (b) *SC*
4. (a) Cut *CT*, *BT* and *AT* has value 21.
   (b) There is a flow of 21.
5. $k = 15$

Exercise 2B (page 24)

1.

2.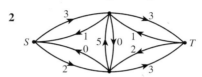

3. (a)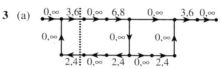
   (b) The cut shown above has value $6 - 2 = 4$.
   The maximum flow is 4.
   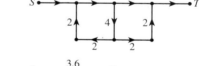

4.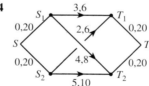

5. 13 million

6. (a) Put all flows equal to the minimum capacities.
   (b) Augment by 4 *S-A-B-T* and 3 *S-C-T*. Total flow = 18.
   (c) Cut *CT* and *BT*.

7.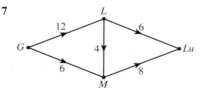
   Maximum flow = 14.

8 (a) $8 \le$ flow ≤ 10
(b) $x \le 10$, $y \ge 8$, $y \ge x$

9 (a) 40 000
(b) Cut M-E and Le-E.

Miscellaneous exercise 2 (page 26)

1 (a) 560, 360
(b) The maximum flow is at most 360. (It is actually 230.)

2 (a) Cut AC, BC, DT has value 22.
(b) The established flow is only 19 whereas the maximum flow must be 22.
(c) S-A-D-B-C-T of extra flow 3

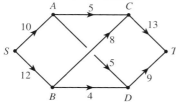

3 (a)

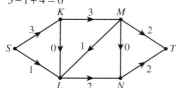

Other solutions are possible.
(b) 11 litres per second

4 (a) $3 - 1 + 4 = 6$
(b)

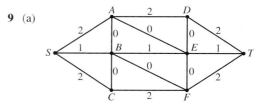

(c) $4 \le$ maximum flow ≤ 6

5 (a) 32
(b) Flow ≤ 32.
(d) UW, VW, XW, XY, XZ (plus possibly others)
These edges form a cut of value of 23, so the maximum flow is 23.

6 (a) 5. Cut SA, BA and BT.
By the maximum flow/minimum cut theorem, the flow of 5 is maximal.
(b) These represent 'flow in = flow out' for vertices A and B.
(c) The objective function is the flow out of S. Alternatively use the flow into T, which is $x_{AT} + x_{BT}$.

7 (a) 5, 4
(b)

SA, SB	S	A, B, C, T | 5
SA, BC	S, B	A, C, T | 4
AT, AC, SB	S, A	B, C, T | 5
AT, AC, BC	S, A, B	C, T | 4
SA, AC, CT	S, B, C	A, T | 6
AT, CT	S, A, B, C	T | 5

(c) 4000 vehicles per hour

8 (a) Join S_1, S_2 and S_3 to a single source S and T_1 and T_2 to a single sink T, with capacities of, say, 60.
(b) 40
(c)

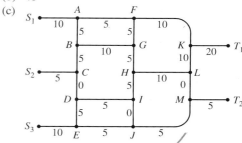

Other solutions are possible.
(d) (i) 30, by augmenting by 5 on S_2-C, C-D, D-I, I-J, J-M, M-T_2.
(ii) AF, AB, BC, CD, S_3E is a cut of 30, so by the maximum flow/minimum cut theorem the maximum is 30.

9 (a)

Other solutions are possible.
(b) Augment by 1 on S-B, B-E, E-T and 1 on S-C, C-B, B-A, A-E, E-D, D-T; 7.
(c) This is the maximum flow, because there is a cut of 7, AD, DE, ET, FT.

10 (a) Replace C by an edge from C_1 to C_2 with maximum capacity 10 and minimum 5. AC and BC are replaced by AC_1 and BC_1 and CD and CE are replaced by C_2D and C_2E.
(b) $4 - 2 + 4 + 6 + 2 = 14$
(c) The maximum flow is less than or equal to 14. (It is actually 11.)

11 (a) Cut = 8 − 1 + 4 = 11
   (b) Flows of 5 on *SA*, 2 on *SB*, 4 on *SC*, 1 on *CB*, 3 on *BA*, 8 on *AT*, 3 on *CT*; as the flow is equal to the cut, it is a maximum by the maximum flow/minimum cut theorem.
   (c) Augment *SABT* by 2 and *SBT* by 1. The new network has flows of 7 on *SA*, 3 on *SB*, 4 on *SC*, 1 on *CB*, 1 on *BA*, 8 on *AT*, 3 on *BT*, 3 on *CT*.

3 Critical path analysis

In this chapter there is not room to give the answers in the form of the examples in the text.

Exercise 3A (page 35)

1 (a)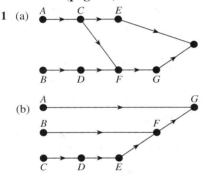
   (b)

2 $A(0|4)$, $B(0|5)$, $C(3|11)$, $D(3|10)$, $E(5|11)$, $F(5|10)$, $G(11|18)$, $H(9|18)$

3 (a) $A(0|4)$, $B(0|10)$, $C(4|13)$, $D(8|13)$, $E(13|20)$, $F(13|15)$, $G(15|20)$
   (b) $A(0|10)$, $B(0|9.5)$, $C(0|8)$, $D(2|8\tfrac{1}{2})$, $E(2\tfrac{1}{2}|9\tfrac{1}{2})$, $F(3\tfrac{1}{2}|10)$, $G(10|10)$

4 17

5 (a)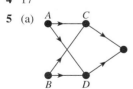

   (b)
Activity	A	B	C	D
Early	0	0	3	3
Late	0	1	6	3

6 (a)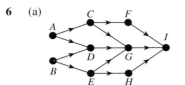

   (b)
Activity	A	B	C	D	E	F	G	H	I
Early	0	0	2	2	2	3	5	5	9
Late	1	0	4	3	2	8	5	6	9

Exercise 3B (page 38)

1 *ABHI*

2 Activity *I* because it is the only one which is critical.

3 (a) $A(0|3)$, $B(3|10)$, $C(0|7)$, $D(7|10)$, $X(3|7)$, $END(10|10)$
   (b) $A(0|3)$, $B(3|9)$, $C(0|6)$, $D(6|9)$, $X(3|6)$, $END(9|9)$
       $t \geq 3$

4 (a)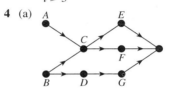

   (b)
Activity	A	B	C	D	E	F	G
Early	0	0	5	5	13	13	8
Late	3	0	5	9	13	14	12

   (c) *BCE*, 15 days
   (d) *BDG*

5 (a)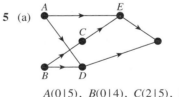

   $A(0|5)$, $B(0|4)$, $C(2|5)$, $D(5|8)$, $E(5|8)$, $END(8|8)$
   (b) 8 days; *A*, *E*

6 (a)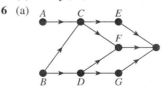

   (b) *BCE*, 11 weeks
   (c) *F*, by 3 weeks

7 (a) *AD*, 8 days (b) 5 days

Exercise 3C (page 42)

1 (a) 20 hours; B, C, E, G
   (b)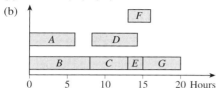
   (c) 3 people
       Delay the start of F until D has finished.

2 (a)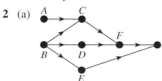
   (b) BE, 6 days
   (c)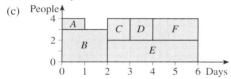

3 (a) AFG, 30 days
       Start times:

Activity	A	B	C	D	E	F	G
Early	0	0	3	3	12	12	24
Late	0	3	8	6	18	12	24

       Finish times:

Activity	A	B	C	D	E	F	G
Early	12	3	7	9	18	24	30
Late	12	6	12	12	24	24	30

   (b) 3
   (c) 1 day. Run C and D one after the other.

4 (a)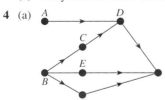
       BE, 26 days
   (b) People graph
   (c) Delay F to start after activity C.

Miscellaneous exercise 3 (page 44)

1 (a)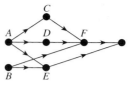
   (b) 10 days; B, E, F
   (c) £1250

2 (a)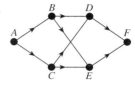
       The early and late times $A(0|5)$, $B(5|7)$, $C(5|7)$, $D(7|11)$, $E(7|11)$, $F(11|12)$
   (b) Start times

Activity	A	B	C	D	E	F
Early	0	5	5	7	7	11
Late	0	5	6	9	7	11

   (c) ABEF, 12 hours
   (d) E now has an effective duration of 5 hours. The critical path is 1 hour longer.

3 (a) [network diagram]
   (b,c)

Activity	A	B	C	D	E	F	G	H	I
Early start	0	1	1	5	5	5	8	14	17
Late finish	1	5	11	11	14	17	17	17	19

   (d) ABEHI, 19 days
   (e)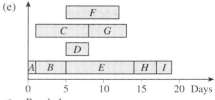
   (f) People graph
   (g) For example, delay the start of F for 3 days.

4 (a)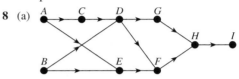

The early and late times reading from the left are (0|0), (3|4), (4|4), (5|6), (7|7), (7|11), (11|12), (14|14), (14|16), (20|20), (23|23).

(b,c)
Act.	A	B	C	D	E	F	G	H	I	J	K
E. s.	0	0	3	4	7	5	7	14	14	11	20
L. f.	4	4	6	7	16	12	14	20	20	20	23

(d) *BDGIK*, 23 weeks

(e) (i) Speed up *I* by 2 weeks since it is on the critical path. Leave *E* as it is not on the critical path. Speed up *J* since it is now on the critical path. Only 1 week is relevant.
(ii) 21 weeks
(iii) £11,000

5 (a)

(b)
Activity	A	B	C	D	E	F	G	H	I	J
Early start	0	0	7	7	5	5	10	12	13	22
Late start	0	4	7	9	13	9	18	16	13	22

(c) The slack times are *A* 0, *B* 4, *C* 0, *D* 2, *E* 8, *F* 4, *G* 8, *H* 4, *I* 0, *J* 0; *ACIJ*, 25 days

6 (a)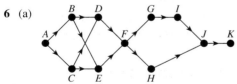

(b)
Act.	A	B	C	D	E	F	G	H	I	J	K
E. s.	0	3	3	9	9	15	17	17	18	20	26
L. s.	0	3	4	11	9	15	18	17	19	20	26

(c) *ABEFHJK*, 27 days

(d) *G* and *I* become critical instead of *H*, saving 1 day.

7 (a)

(b)
Activity	A	B	C	D	E	F
Early start	0	0	3	2	4	8
Late start	0	1	3	3	4	8

(c) *A, C, E, F*; 13 days

(d) Six people will be required for each of four days between days 2 and 8. Four of these people will do *D*, and the remainder whichever of *A*, *C* and *E* is going on in parallel.

8 (a)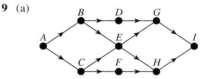

(b)
Activity	A	B	C	D	E	F	G	H	I
Early start	0	0	45	57	45	59	59	62	63
Late start	0	39	45	57	55	59	61	62	63

(c) *ACDFHI*, 66 days

(d) 10 days

9 (a)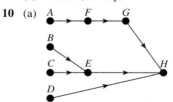

(b)
Activity	A	B	C	D	E	F	G	H	I
Early start	0	5	5	7	9	9	13	13	16
Late start	0	7	5	9	9	10	14	13	16

(c) *ACEHI*, 20 days

10 (a)

(b) 22 minutes
(c) *D, H*
(d) 32 minutes

11 (a) (i, ii, iii)

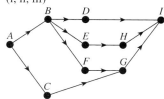

Activity	A	B	C	D	E	F	G	H	I
Early start	0	2	2	8	8	8	14	24	28
Duration	2	6	12	12	16	4	10	4	2
Late finish	2	8	18	28	24	18	28	28	30

(iv) *ABEHI*, 30 days

(b)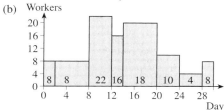

(c) Move *D* back 4 days to 8, or *F* and *G* back 4 days.

12 (a, b, c)

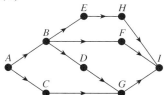

Activity	A	B	C	D	E	F	G	H	I
Early start	0	1	1	5	5	5	8	14	17
Duration	1	4	7	3	9	7	6	3	2
Late finish	1	5	11	11	14	17	17	17	19

(d) *ABEHI*, 19 days

(e)

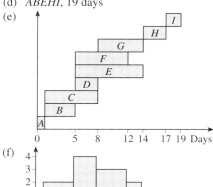

(g) For example,
worker 1, *ABEHI*;
worker 2, *CDG*;
worker 3, *F*.

13 (a, b, c)

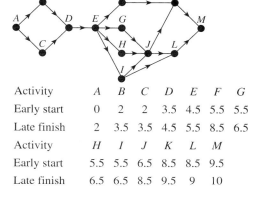

Activity	A	B	C	D	E	F	G
Early start	0	2	2	3.5	4.5	5.5	5.5
Late finish	2	3.5	3.5	4.5	5.5	8.5	6.5

Activity	H	I	J	K	L	M
Early start	5.5	5.5	6.5	8.5	8.5	9.5
Late finish	6.5	6.5	8.5	9.5	9	10

(d) *ACDEHIJKM*, 10 days

(e)

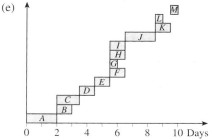

(f) Electrician, *B*, *G*- $\frac{3}{4}$ day

Plumber *C*, (*I*)- $\frac{3}{8}$ day

Joiner (*H*), *J*- $\frac{2}{5}$ day

Use joiner – electrics not critical;
new completion time is $9\frac{3}{5}$ days.

4 Dynamic programming

Exercise 4 (page 59)

1 *A*(5,1), *B*(3,1), *C*(2,1), *D*(0,1), *E*(1,1), *F*(4,1)

2 *AEBCD*, 6 lights

3 *AGCE*, 17 miles

4 (a) There are negative weights.
(b) *AFBCED*, 7

5 From *A*, *ADGJMN*, 13
From *B*, *BDGJMN*, 12
From *C*, *CHJMN*, 10

Miscellaneous exercise 4 (page 60)

1 (a)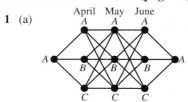

(b) £186

(c) April-*C*, May-*C*, June-*A*; £154

2 (a) Dijkstra's algorithm is of quadratic order compared to the cubic order of the dynamic programming solution. Dijkstra's algorithm will therefore be much quicker for larger problems.

(b) Dijkstra's algorithm cannot be used if some edges have negative weights whereas dynamic programming can.

3

Stage	State	Act.	Value	Minimum
1	1	1	8	←8
	2	1	6	←6
	3	1	10	←10
2	1	1	11 + 8 = 19	
		2	6 + 6 = 12	←12
	2	1	9 + 6 = 15	
		2	4 + 10 = 14	←14
3	1	1	3 + 12 = 15	←15
		2	2 + 14 = 16	
	2	1	13 + 14 = 27	
		2	5 + 10 = 15	←15
4	1	1	6 + 15 = 21	←21
		1	8 + 15 = 23	

SABET, 21

4 (a)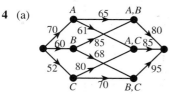

(b) *A, B, C*; £21,500,000

5 (a)

Stage	State	Act.	Value	Minimum
1	1	1	13	←13
	2	1	7	←7
2	1	1	25 + 13 = 38	
		2	28 + 7 = 35	←35
	2	1	19 + 13 = 32	
		2	22 + 7 = 29	←29
	3	1	37 + 13 = 50	
		2	22 + 7 = 29	←29
3	1	1	10 + 35 = 45	
		2	17 + 29 = 46	
		3	13 + 29 = 42	←42

(b) *SCET*, 42

(c) An advantage is that dynamic programming can deal with negative numbers on the edges; a disadvantage is that it has higher polynomial order and so takes longer for large problems.

6 13 cakes, costing £78 must be baked. Remove this from the table, and add later.

Day	State	Make	Store	D. cost	Total	
Fri.	0	2	0	2.5	2.5	←
	1	1	0	2.5	2.5	←
	2	0	0	0	0	←
Thu.	0	4	1	3	5.5	
		3	0	2.5	5	←
	1	4	2	3.5	3.5	←
		3	1	3	5.5	
		2	0	2.5	5	
	2	3	2	3.5	3.5	←
		2	1	3	5.5	
		1	0	2.5	5	
	3	2	2	3.5	3.5	
		1	1	3	5.5	
		0	0	0	2.5	←
Wed.	1	4	0	2.5	7.5	←
	2	4	1	3	6.5	←
		3	0	2.5	7.5	
	3	4	2	3.5	7	
		3	1	3	6.5	←
		2	0	2.5	7.5	
Tue.	0	4	2	3.5	10	←
		3	1	3	10.5	
	1	4	3	4	10.5	
		3	2	3.5	10	←
		2	1	3	10.5	
	2	3	3	4	10.5	
		2	2	3.5	10	←
		1	1	3	10.5	
	3	2	3	4	10.5	
		1	2	3.5	10	
		0	1	0.5	8	←
Mon.	0	4	3	4	12	←
		3	2	3.5	13.5	
		2	1	3	13	
		1	0	2.5	12.5	

Make 4, 0, 4, 3, 2, storing 3, 1, 0, 0, 0, costing £90.

7

Stage	State	Act.	Value	Maximum
1	1	0	5	← 5
	2	1	3	← 3
2	1	0	6 + 5 = 11	
		1	2 + 3 = 5	← 5
	2	0	3 + 5 = 8	← 8
		1	6 + 3 = 9	
3	1	0	3 + 5 = 8	← 8
		1	3 + 8 = 11	
	2	0	5 + 5 = 10	← 10
		1	3 + 8 = 11	
4	1	0	5 + 8 = 13	← 13
		1	4 + 10 = 14	

The route is $(4,1), (3,1), (2,1), (1,2), (0,1)$; 13.

8

Stage	State	Act.	Value	Maximum
1	1	1	3	← 3
	2	1	4	← 4
	3	1	5	← 5
2	1	1	6 + 3 = 9	← 9
	2	1	5 + 3 = 8	
		2	5 + 4 = 9	
		3	5 + 5 = 10	← 10
	3	1	6 + 4 = 10	
		2	6 + 5 = 11	← 11
3	1	1	4 + 9 = 13	
		2	4 + 10 = 14	← 14
	2	1	3 + 9 = 12	
		2	3 + 10 = 13	← 13
	3	1	2 + 10 = 12	
		2	2 + 11 = 13	← 13

Visit $(0,1), (1,3), (2,2), (3,1)$; £14,000

9

Stage	State	Act.	Value	Maximum
1	A	1	11	← 11
	B	1	9	← 9
	C	1	7	← 7
2	A & B	1	10 + 11 = 21	← 21
		2	14 + 9 = 23	
	A & C	1	9 + 11 = 20	
		2	12 + 7 = 19	← 19
	B & C	1	10 + 9 = 19	← 19
		2	12 + 7 = 19	← 19
3	A & B & C	1	12 + 21 = 33	
		2	13 + 19 = 32	← 32
		3	14 + 19 = 33	

C, A, B; £32,000

5 The Simplex algorithm

Exercise 5A (page 70)

1 (a)
$$\begin{array}{ccccc|c} 1 & -2 & -3 & 0 & 0 & 0 \\ 0 & 1 & 2 & 1 & 0 & 6 \\ 0 & 1 & 1 & 0 & 1 & 5 \\ \hline 1 & 0 & -1 & 0 & 2 & 10 \\ 0 & 0 & 1 & 1 & -1 & 1 \\ 0 & 1 & 1 & 0 & 1 & 5 \\ \hline 1 & 0 & 0 & 1 & 1 & 11 \\ 0 & 0 & 1 & 1 & -1 & 1 \\ 0 & 1 & 0 & -1 & 2 & 4 \end{array}$$
Maximum 11 at $(4,1)$.

(b)
$$\begin{array}{ccccc|c} 1 & -2 & -1 & 0 & 0 & 0 \\ 0 & 1 & 2 & 1 & 0 & 6 \\ 0 & 1 & 1 & 0 & 1 & 5 \\ \hline 1 & 0 & 1 & 0 & 2 & 10 \\ 0 & 0 & 1 & 1 & -1 & 1 \\ 0 & 1 & 1 & 0 & 1 & 5 \end{array}$$
Maximum 10 when $y = 0$ and $t = 0$. Since $x + y + t = 5$, $x = 5$, the maximum is 10 at $(5,0)$.

2
$$\begin{array}{ccccc|c} 1 & -1 & -4 & 0 & 0 & 0 \\ 0 & 1 & 3 & 1 & 0 & 15 \\ 0 & 2 & 1 & 0 & 1 & 12 \\ \hline 1 & \frac{1}{3} & 0 & \frac{4}{3} & 0 & 20 \\ 0 & \frac{1}{3} & 1 & \frac{1}{3} & 0 & 5 \\ 0 & \frac{5}{3} & 0 & -\frac{1}{3} & 1 & 7 \end{array}$$
Maximum of 20 at $(0,5)$

3
$$\begin{array}{cccccc|c} 1 & -5 & -3 & 8 & 0 & 0 & 0 \\ 0 & 1 & 0 & -1 & 1 & 0 & 0 & 1 \\ 0 & 1 & 1 & 0 & 0 & 1 & 0 & 2 \\ 0 & 3 & 2 & -4 & 0 & 0 & 1 & 6 \\ \hline 1 & 0 & -3 & 3 & 5 & 0 & 0 & 5 \\ 0 & 1 & 0 & -1 & 1 & 0 & 0 & 1 \\ 0 & 0 & 1 & 1 & -1 & 1 & 0 & 1 \\ 0 & 0 & 2 & -1 & -3 & 0 & 1 & 3 \\ \hline 1 & 0 & 0 & 6 & 2 & 3 & 0 & 8 \\ 0 & 1 & 0 & -1 & 1 & 0 & 0 & 1 \\ 0 & 0 & 1 & 1 & -1 & 1 & 0 & 1 \\ 0 & 0 & 0 & -3 & -1 & -2 & 1 & 1 \end{array}$$

The first of the last four rows tells you that the maximum of 8 occurs when $z = 0$. The other rows tell you that $x = y = 1$. Thus the maximum of 8 occurs at $(1,1,0)$.

4 (a) 10 of A and 15 of B

(b) Two assumptions are that all the items can be sold, and that there are no contracts for C.

5 68; 40 of Smart Savings, 20 of Capital Investor and 8 of Money Monthly

6 Minimum -4 at $(2,0,3)$

Exercise 5B (page 72)

1 (a) $6x+5y+3z$; $7x+7y+4z \leq 23$,
$5x+6y+2z \leq 16$, $4x+8y-2z \leq 13$;
x, y, z; r, s, t

(b) $19\frac{1}{2}$; $x=3$, $y=0$, $z=\frac{1}{2}$

2 (a) 90 at $(10, 30)$

(b) 10 at $(0, 5)$

(c) $(0, 0)$ is not in the feasible region.

(d)

Graph: Max vs c, with points at (5,10), (40,80), (68,108); labelled "c < 5 is impossible"

3 Produce 2500 Supremo 64s, 250 Supremo 128s and 250 Supremo 256s. This assumes that all models are sold, and that there are no extra costs and overheads.

4 (a)
1	−3	−3	−2	0	0	0
0	3	7	2	1	0	15
0	2	4	1	0	1	8

(b)
1	0	3	−0.5	0	1.5	12
0	0	1	0.5	1	−1.5	3
0	1	2	0.5	0	0.5	4

(c) $P=12$, $x=4$, $y=z=0$; this is not optimal because there is a negative number in the z-column of the top row.

(d) $P=15$, $x=1$, $y=0$, $z=6$

5 $v = n - m$. This is usually, but not always, true. It depends on the linear constraints. A proof depends upon knowledge from matrix algebra.

Miscellaneous exercise 5 (page 73)

1 (a)
1	−4	−5	−3	0	0	0
0	8	5	2	1	0	4
0	1	2	3	0	1	1

(b) $P=2$, $x=\frac{1}{2}$, $y=z=0$

(c) $P = 2\frac{10}{11}$, $x = \frac{3}{11}$, $y = \frac{4}{11}$, $z = 0$; top row of tableau has no negative numbers.

2 (a)
1	−4000	−6000	0	0	0	0
0	1	1	1	0	0	5000
0	−1	4	0	1	0	0
0	1	3	0	0	1	6000

(b)
1	0	−2000	4000	0	0	20 million
0	1	1	1	0	0	5000
0	0	5	1	1	0	5000
0	0	2	−1	0	1	1000

This shows that $P = 20$ million when $x = 5000$ and $y = 0$.

(c) The optimal solution has not been reached because the element at the top of the y-column is negative.

3 (a) Letting x, y and z be the number of Brainy, Superbrainy and Superbrainy X robots made, the problem is to maximise $P = 2400x + 3000y + 3200z$ subject to the constraints $6x + 3y + z \leq 60$, $3x + 4y + 5z \leq 60$ and $x, y, z \geq 0$ are integers.

(b)
1	−2400	−3000	−3200	0	0	0
0	6	3	1	1	0	60
0	3	4	5	0	1	60

(c)
1	0	−1800	−2800	400	0	24000
0	1	0.5	$\frac{1}{6}$	$\frac{1}{6}$	0	10
0	0	2.5	4.5	−0.5	1	30

The profit is £24,000 at $x = 10$, $y = z = 0$, but the optimum has not yet been reached.

4 (a)
1	−3	−4	−5	0	0	0
0	2	8	5	1	0	3
0	9	3	6	0	1	2

1	−3	−4	−5	0	0	0
0	2	8	5	1	0	3
0	9	3	6	0	1	2
1	$4\frac{1}{2}$	$-1\frac{1}{2}$	0	0	$\frac{5}{6}$	$1\frac{2}{3}$
0	$-5\frac{1}{2}$	$5\frac{1}{2}$	0	1	$-\frac{5}{6}$	$1\frac{1}{3}$
0	$1\frac{1}{2}$	$\frac{1}{2}$	1	0	$\frac{1}{6}$	$\frac{1}{3}$
1	3	0	0	$\frac{3}{11}$	$\frac{20}{33}$	$2\frac{1}{33}$
0	−1	1	0	$\frac{2}{11}$	$-\frac{5}{33}$	$\frac{8}{33}$
0	2	0	1	$-\frac{1}{11}$	$\frac{8}{33}$	$\frac{7}{33}$

(b) After 1 iteration, $x = y = 0$, $z = \frac{1}{3}$, $P = 1\frac{2}{3}$.
After 2 iterations, $x = 0$, $y = \frac{8}{33}$, $z = \frac{7}{33}$, $P = 2\frac{1}{33}$.

(c) There are no negative entries in the top row, so this is optimal.

5 (a, b)
1	−2	−1	0	0	0	0
0	1	1	1	0	0	7
0	1	2	0	1	0	10
0	2	3	0	0	1	16
1	0	1	2	0	0	14
0	1	1	1	0	0	7
0	0	1	−1	1	0	3
0	0	1	−2	0	1	2

(c) $x = 7$, $y = 0$, $P = 14$

(d) The top row in the tableau now has no negative signs, so the optimum has been reached.

6 (a, b)

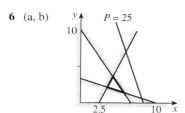

(c) $x = 5.7, y = 1.4$

(d)
1	-8	-6	0	0	900
0	-7	1	1	0	0
0	7	9	0	1	350
1	0	$\frac{30}{7}$	0	$\frac{8}{7}$	1300
0	0	10	1	1	350
0	1	$\frac{9}{7}$	0	$\frac{1}{7}$	50

Optimum $R = 1300$, when $Y = 0$ and $X = 50$.

(e) Optimum $P = \frac{130}{7}$, when $x = \frac{40}{7}$, $y = \frac{10}{7}$.

6 Game theory

Exercise 6A (page 80)

1 $\begin{pmatrix} -3 & 2 & -1 & 2 \\ -1 & 1 & 0 & -1 \\ 2 & -4 & -2 & 3 \end{pmatrix}$

2 (a) $A1, B2, 4$ (b) $A2, B2, 3$
(c) $A1, B4, -2$ (d) $A1, B2, -3$

3 (a) Yes, value 4 (b) Yes, value 3
(c) No (d) No

4 (a) Yes (b) No
(c) Yes (d) Yes

5 $A3, B3$. The game has a stable solution so neither player can improve their pay-off.

6 (a) It is dominated by strategy number 2.
(b) Y
(c) $\begin{pmatrix} 0.2 & 0.1 \\ 0.3 & 0.6 \end{pmatrix}$, United 2, City X
(d) Strategy number 2 as the solution is stable.

Exercise 6B (page 86)

1 (a) $X: 1+2p$; $Y: 3-5p$; $Z: 2-p$
(b) $\frac{2}{7}$
(c) Ann $\frac{11}{7}$, Ben $-\frac{11}{7}$
(d) Strategy Z

2 (a) $3+2p, 4-2p$
(b) $2+3q, 4-q$
(c) $\frac{7}{2}; \frac{1}{4}, \frac{1}{2}$

3 (a) $\frac{17}{5}$ (b) $\frac{5}{3}$ (c) $\frac{5}{7}$

4 (a) $\begin{pmatrix} 6 & 2 \\ 1 & 5 \end{pmatrix}$ (b) $\frac{7}{2}$

5 Mixture is dominated by New.
(a) $8-7p$, $5-2p$, $3+5p$
(b)

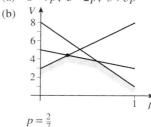

$p = \frac{2}{7}$

Exercise 6C (page 90)

1 Maximise v,
subject to $v \leqslant 4p + 3q$,
$v \leqslant p + 5q$,
$p + q \leqslant 1$,
$p, q \geqslant 0$.

P	v	p	q	r	s	t	
1	-1	0	0	0	0	0	0
0	0	1	1	1	0	0	1
0	1	-4	-3	0	1	0	0
0	1	-1	-5	0	0	1	0
1	0	-4	-3	0	1	0	0
0	0	1	1	1	0	0	1
0	1	-4	-3	0	1	0	0
0	0	3	-2	0	-1	1	0
1	0	0	$-\frac{17}{3}$	0	$-\frac{1}{3}$	$\frac{4}{3}$	0
0	0	0	$\frac{5}{3}$	1	$\frac{1}{3}$	$-\frac{1}{3}$	1
0	1	0	$-\frac{17}{3}$	0	$-\frac{1}{3}$	$\frac{4}{3}$	0
0	0	1	$-\frac{2}{3}$	0	$-\frac{1}{3}$	$\frac{1}{3}$	0
1	0	0	0	$\frac{17}{5}$	$\frac{4}{5}$	$\frac{1}{5}$	$\frac{17}{5}$
0	0	0	1	$\frac{3}{5}$	$\frac{1}{5}$	$-\frac{1}{5}$	$\frac{3}{5}$
0	1	0	0	$\frac{17}{5}$	$\frac{4}{5}$	$\frac{1}{5}$	$\frac{17}{5}$
0	0	1	0	$\frac{2}{5}$	$-\frac{1}{5}$	$\frac{1}{5}$	$\frac{2}{5}$

$v = \frac{17}{5}$; $p = \frac{2}{5}$, $q = \frac{3}{5}$

2 Maximise $v - 2$,
subject to $v \leqslant p + 4q$,
$v \leqslant 5p + q$,
$p + q \leqslant 1$,
$p, q \geqslant 0$.

$v = \frac{5}{7}$

3 Maximise v,
subject to $v \leqslant 3p + 4q + 2r$,
$v \leqslant 2p + 5q + 4r$,
$v \leqslant 5p + 2q + 4r$,
$p + q + r \leqslant 1$,
$p, q, r \geqslant 0$.

Miscellaneous exercise 6 (page 90)

1. max(row min) = min(col max) = 1
   Both play option 2.

2. (a) $1-p-2q$, $2p+q-1$, $q-p$
   (b) By choosing $p = q = 1-p-q = \frac{1}{3}$
   (c) 0. Each player can guarantee this by choosing their options with equal probabilities.

3. (a) max(row min) = -2 and min(col max) = 1; these are unequal so the game does not have a stable solution.
   (b) (i) 1   (ii) -0.8 (iii) 0.1
   (c) A plays 1 with probability $\frac{3}{8}$; the pay-off is $-\frac{1}{8}$.

4. (a) Roland plays S, gain ≥ -1, and Colleen plays S, loss ≤ 3. As these outcomes are unequal, the game does not have a stable solution.
   (b) (i) T (ii) S
   (c) $5p-2$
   (d)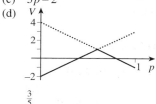

5. (a) Atomic energy is dominated by Bio-gas, and Windy is dominated by Wet.
   (b) Bio-gas and Wet; max(row min) = 2 and min(col max) = 4; these are unequal so the game does not have a stable solution.
   (c) $1+3p$
   (d) $\frac{2}{3}$, 3

6. (a) P and B; max(row min) = -1 and min(col max) = 1; these are unequal so the game does not have a stable solution.
   (b) (i) P (ii) A
   (c) $3-4p$
   (d) $\frac{7}{9}$, $\frac{1}{9}$

7. (a) Rowena: B, guarantees ≥ 3, Colin: Z, guarantees ≤ 3; these are equal, so the game has a stable solution.
   (b) There is no advantage to either player in varying from play-safe strategy.
   (c) X is dominated by Z;
   C is dominated by B or A is dominated by B.

	Y	Z			Y	Z
A	-1	2	or	B	6	3
B	6	3		C	2	-2

   (d) $3+3p$
   (e) 0, -3

8. (a) Zero-sum game means that the gain of one player added to the gain of the other player is zero.
   (b) Richard: A, gain ≥ -2;
   Carol: Z, loss ≤ 1;
   these are unequal so the game does not have a stable solution.
   (c) Richard will change to C.
   (d) Carol will not change.

9. (a) Rose: Hide, gain ≥ 3;
   Computer: Shoot, loss ≤ 4; these are unequal so the game does not have a stable solution.
   (b) It is dominated by both of the other strategies.
   (c) Track, because it is dominated by Shoot.
   (d) Computer chooses Shoot, gain $5p-1$; computer chooses Fight, gain $5-2p$.
   (e) $\frac{6}{7}$

10. (a) Elf is dominated by Dragon.
    (b) Crystal
    (c) New matrix is

	D	F
A	0.4	0.3
B	0.5	0.8

    Richard: Bronze, expected gain 0.5;
    Computer: Dragon, expected gain to Richard 0.5.
    (d) The game has a stable solution, so Richard should always play Bronze.

11. (a) max(row min) = 1 and min(col max) = 3; these are unequal so the game does not have a stable solution.
    (b) For B, Z is always better than X.
    (c) B plays Y with probability $\frac{7}{15}$ and Z with probability $\frac{8}{15}$.

Revision exercise
(page 95)

1. (a)

Activity	A	B	C	D	E	F	G	H	I
Early start	0	0	2	3	3	3	5	5	8
Late start	1	0	2	5	3	4	6	5	8
Early finish	1	2	3	4	5	4	7	8	9
Late finish	2	2	3	6	5	5	8	8	9
Float	1	0	0	2	0	1	1	0	0

(b) 9 days
(c) B, C, E, H, I
(d) It need have no effect because G has a float of 1 day.

2 (a) For example, replace all scores by '8 – score'.
(b) Add a row of equal scores.
(c) L-W, M-X, N-Y or L-Z, M-X, N-W

3 (a)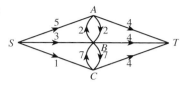

(b) A cut is any continuous line which separates S from T. It must not pass through any vertices.
(c) 8
(d)

X	Y	Value
S	T, A, B, C	9
S, A	T, B, C	10
S, B	T, A, C	19
S, C	T, A, B	19
S, A, B	T, C	16
S, A, C	T, B	17
S, B, C	T, A	15
S, A, B, C	T	12

(e) The maximum possible flow is 9.

4 (a) SDCT, SDBCT, SDBT, SADCT, SADBCT, SADBT, SABCT, SABT
(b) C_1: 340; C_2: 250. The maximum flow is at most 250 people per hour.
(c) 200 people per hour

5 (a)

Activity	A	B	C	D	E	F	G	H	I	J
Early start	0	2	2	2	2	4	3	7	8	18
Late start	0	13	10	2	2	4	3	15	8	18
Early finish	2	4	7	4	3	8	8	10	18	22
Late finish	2	15	15	4	3	8	8	18	18	22

(b) A, D, E, F, G, I, J
(c) Delay the start of C until day 8. Delay the start of H until day 13.

6 1-A, 2-B, 3-C, 4-D or 1-C, 2-B, 3-A, 4-D or 1-C, 2-B, 3-D, 4-A

7 (3,1), (2,2), (1,2), (0,1); 7

8 (a, b)

P	x	y	z	r	s	
1	−4	−2	−3	0	0	0
0	3	7	2	1	0	15
0	2	4	1	0	1	8
1	0	6	−1	0	2	16
0	0	1	$\frac{1}{2}$	1	$-\frac{3}{2}$	3
0	1	2	$\frac{1}{2}$	0	$\frac{1}{2}$	4

(c) $x = 4$, $y = z = 0$, $P = 16$; the negative number in the z-column means that the solution is not optimal.

9 (a) Maximise $v - 6$,
subject to $v \leq 8p_1 + 2p_2 + p_3$,
$v \leq 9p_1 + 5p_2 + 6p_3$,
$v \leq 4p_1 + 5p_2 + 7p_3$,
$p_1 + p_2 + p_3 \leq 1$,
$p_1, p_2, p_3 \geq 0$.

(b)

P	v	p_1	p_2	p_3	r	s	t	u	
1	−1	0	0	0	0	0	0	0	−6
0	1	−8	−2	−1	1	0	0	0	0
0	1	−9	−5	−6	0	1	0	0	0
0	1	−4	−5	−7	0	0	1	0	0
0	0	1	1	1	0	0	0	1	1

10 (a) His row is dominated by Minnie's.
(b) 5, 6, 5.5
(c) $6 - 2p$, $5 + 2p$, $7 - 3p$
(d)

				Min
Minnie	4	7	4	4
Max	6	5	7	5 ←

Max should play.

11 SAFT; profit of £300.

Mock examinations

Mock examination 1 (page 100)

1 (a)

		Value
S	T, A, B, C	21
S, A	T, B, C	28
S, B	T, A, C	23
S, C	T, A, B	21
S, A, B	T, C	24
S, A, C	T, B	28
S, B, C	T, A	19
S, A, B, C	T	20

(b) For example,

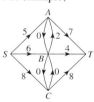

The flow, 19, is equal to the minimum cut, so the flow is maximal.

2

Stage	State	Action	Value		Minimum
1	(1,1)	1	8	←	8
	(1,2)	1	7	←	7
	(1,3)	1	8	←	8
2	(2,1)	1	6 + 8 = 14	←	14
		2	8 + 7 = 15		
	(2,2)	1	7 + 8 = 15		
		2	8 + 7 = 15		
		3	6 + 8 = 14	←	14
	(2,3)	1	8 + 7 = 15	←	15
		2	7 + 8 = 15		
3	(3,1)	1	8 + 14 = 22		
		2	7 + 14 = 21	←	21
		3	8 + 15 = 23		

(3,1), (2,2), (1,3), (0,1); total £21

3 (a)

P	x	y	r	s	t	
1	−3	−2	0	0	0	0
0	1	1	1	0	0	10
0	1	2	0	1	0	15
0	3	1	0	0	1	22
1	0	−1	0	0	1	22
0	0	$\frac{2}{3}$	1	0	$-\frac{1}{3}$	$\frac{8}{3}$
0	0	$\frac{5}{3}$	0	1	$-\frac{1}{3}$	$\frac{23}{3}$
0	1	$\frac{1}{3}$	0	0	$\frac{1}{3}$	$\frac{22}{3}$

(b)

1	0	0	0	$\frac{3}{2}$	0	$\frac{1}{2}$	26
0	0	1	0	$\frac{3}{2}$	0	$-\frac{1}{2}$	4
0	0	0	0	$-\frac{5}{2}$	1	$\frac{1}{2}$	1
0	1	0	0	$-\frac{1}{2}$	0	$\frac{1}{2}$	6

$P = 26$;
$r = 0$ so $x + y = 10$,
$s = 1$ so $x + 2y = 14$,
$t = 0$ so $3x + y = 22$.

4 (a)

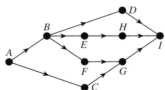

Act.	A	B	C	D	E	F	G	H	I
E. s.	0	10	10	30	30	30	55	95	107
L. f.	10	30	82	107	95	82	107	107	117

(b) 117 minutes, *ABEHI*

(c)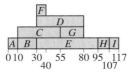

5 (a) Subtract each element from, say, 10.
(b) Add another column of equal numbers, say, 1s.
(c)
```
1 1 0 1 0
0 0 2 1 1
1 1 0 0 0
2 1 1 1 0
1 0 1 0 1
```
Amy-III, Brad-I, Cath-IV, Ed-II

6 (a) It is dominated by *C*.
(b) *L*; it is dominated by *R*.
(c)

	C	R
L	0	2
C	3	1

Ken: *C*, value ⩾ 1; Patrick: *R*, value ⩽ 2; these are unequal so the game does not have a stable solution.

(d) Patrick should choose *C* with probability $\frac{1}{4}$ and *R* with probability $\frac{3}{4}$. Ken should then expect to gain $1\frac{1}{2}$ minutes.

Mock examination 2 (page 103)

1 (a)

P	x	y	r	s	
1	−3	−1	0	0	0
0	1	2	1	0	8
0	2	1	0	1	10

(b)

1	0	$\frac{1}{2}$	0	$\frac{3}{2}$	15
0	0	$\frac{3}{2}$	1	$-\frac{1}{2}$	3
0	1	$\frac{1}{2}$	0	$\frac{1}{2}$	5

(c) $P = 15$; all the elements in the top row are non-negative.

2 (a) There are negative weights.

(b)

Stage	State	Action	Value		Minimum
1	B(1,1)	1	4	←	4
	C(1,2)	1	2	←	2
2	D(2,1)	1	2 + 4 = 6		
		2	5		
		3	2 + 2 = 4	←	4
3	E(3,1)	1	−2 + 4 = 2		
		2	−3 + 4 = 1	←	1
	F(3,2)	1	−2 + 4 = 2		
		2	−1 + 2 = 1	←	1
4	G(4,1)	1	1 + 1 = 2	←	2
		2	−1 + 4 = 3		
		3	2 + 1 = 3		

GEDCA; total 2.

3 *A*-The Galaxy, *B*-Earth Today, *C*-News in Britain, *D*-Daily Echo; total £1640

4 (a) They refer to 'flow in = flow out' for vertices A, B and C respectively.

(b)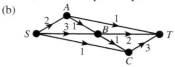

(c) It is 'flow out' for vertex S. Alternatively use 'flow in' for vertex T, $x_{AT} + x_{BT} + x_{CT}$.

(d) For example, cut $\{S, A, B\}$, $\{T, C\}$ has value 5.
For example, flows of 1: SAT, 1: $SABCT$, 1: SCT and 2: SBT have total value 5.
The maximum flow is 5, by the maximum flow/minimum cut theorem.

5 (a)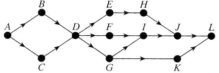

Act.	A	B	C	D	E	F	G	H	I	J	K	L
E. s.	0	2	2	6	9	9	9	11	13	16	13	19
L. f.	2	6	6	9	14	14	13	17	17	19	19	20

20 days; $ABDGKL$

(b) [histogram as shown]

(c) Delay the start of each of E and H by 3 days, and delay the start of J by 1 day.

6 (a) Real Valencia: A, value ≥ 0.2;
Newton Heath: Y, value ≤ 0.4.
These are unequal so the game does not have a stable solution.

(b) Real Valencia should not change as $0.4 > 0.1$.

(c) Newton Heath should change to X to reduce the value to 0.2.

(d) $A: 0.2p + 0.4(1-p) = 0.4 - 0.2p$
$B: 0.5p + 0.1(1-p) = 0.1 + 0.4p$

(e) $p = 0.5$; value 0.3

(f) Choose A with probability $\frac{2}{3}$ and B with probability $\frac{1}{3}$.

Glossary

Action	Edge of a dynamic programming network.
Activity network	Network which represents the splitting up of a major project into activities.
Allocation problem	Matching problem where total cost must be minimised.
Capacities	Weights on edges in network flow problems.
Cascade diagram	Diagram showing possible time slots of all activities involved in a project.
Critical path	Path through an activity network consisting only of critical activities which have no scheduling flexibility.
Cut	Continuous line separating the source (start vertex) from the sink (end vertex).
Dominance	Process of ignoring a particular choice of action in a two-person game when that choice is always inferior to another choice.
Float	Flexibility in scheduling an activity.
$m \times n$ game	Two-person zero-sum game where one player has m possible choices of action and the other player has n possible choices.
Matching	Set of edges of a bipartite graph which have no vertices in common.
Pay-off matrix	Rectangular array of numbers representing outcomes of pairs of decisions.
Resource levelling	Smoothing out of the use of a resource.
Saddle point	Position of a stable solution in a pay-off matrix.
Stable solution	Solution where neither player in a game benefits from changing from a play-safe strategy.
State	Vertex of a dynamic programming network.
Zero-sum game	Game in which the gain for one party is exactly matched by the loss for the other.

Summary of algorithms

Problem	Name of algorithm
Allocation	Hungarian
Maximum flow	Labelling
Critical path analysis	Drawing an activity network
	Early time
	Late time
Linear programming	Simplex method
Minimising	Dynamic programming
Zero-sum games	Play-safe strategy
	Mixed strategy

Index

The page numbers refer to the first mention of each term, or the shaded box if there is one.

action, 53
activity, 30
    critical, 37
activity network, 30
    procedure for drawing, 32
algorithm
    drawing an activity network, 32
    early time, 33
    Hungarian, 6
    labelling procedure, 19
    late time, 34
    Simplex, 67
    solving a $2 \times n$ game, 84
allocation problem, 1
augmenting the elements, 6

Bellman's optimising principle, 55

capacity, 15
    excess, of a network, 18
    lower, of an edge, 22
    of a cut, 16
    restricted, of a vertex, 22
    upper, of an edge, 22
cascade diagram, 39
cost, of an action, 53
critical activity, 37
critical path, 37
critical path analysis, 30
cut, 15
    capacity of, 16
    value of, 16, 22

dominance, 80
dynamic programming, 53

earliest finish time, 35
earliest start time (early time), 32
early time algorithm, 33

float, 37
forward pass, 32

game theory, 76
games
    $2 \times n$, 84
    $m \times n$, 87
    value of, 79
    zero-sum, 76
Gantt diagram, 40
graph of expected pay-off, 82

Hungarian algorithm, 6
    non-square arrays, 9

labelling procedure algorithm, 19
late time algorithm, 34
latest start time (late time), 35

matching, 1
matrix, pay-off, 77
maximum flow, 16
maximum flow/
    minimum cut theorem, 16
minimum cut, 16
mixed strategy, 81

negative edge weights, 53
network of possible increases, 18
non-square arrays
    for the Hungarian algorithm, 9

optimisation problems, 53
outcome, of pay-off matrix, 77

pay-off matrix, 77
play-safe strategy, 77
potential backflow, 18

resource histogram, 41
resource levelling, 41
reverse pass, 34

saddle point, 79
saturated edge, 15
shortest path, 71
Simplex algorithm, 67
Simplex method, 64
Simplex tableau, 65
sink, 15
slack variable, 64
solving a $2 \times n$ game, 84
source, 15
stable solution, 79
stage, 54
stage variable, 53
state, 53
state variable, 54
strategy
    mixed, 81
    play-safe, 77
    sub-optimal, 55
sub-optimal strategy, 55
supersink, 21
supersource, 21

tabulation, dynamic programming, 55
two-person zero-sum game, 77

value
    of a cut, 16
    of a game, 79
variable
    slack, 64
    stage, 53
    state, 54

zero-sum game, 76